Security Informatics and Law Enforcement

Series Editor
Babak Akhgar
CENTRIC (Centre of Excellence in Terrorism, Resilience, Intelligence and Organised Crime Research)
Sheffield Hallam University
Sheffield, UK

The primary objective of this book series is to explore contemporary issues related to law enforcement agencies, security services and industries dealing with security related challenges (e.g., government organizations, financial sector insurance companies and internet service providers) from an engineering and computer science perspective. Each book in the series provides a handbook style practical guide to one of the following security challenges:

Cyber Crime – Focuses on new and evolving forms of crimes. Books describe the current status of cybercrime and cyber terrorism developments, security requirements and practices.

Big Data Analytics, Situational Awareness and OSINT – Provides unique insight for computer scientists as well as practitioners in security and policing domains on big data possibilities and challenges for the security domain, current and best practices as well as recommendations.

Serious Games – Provides an introduction into the use of serious games for training in the security domain, including advise for designers/programmers, trainers and strategic decision makers.

Social Media in Crisis Management – explores how social media enables citizens to empower themselves during a crisis, from terrorism, public disorder, and natural disasters.

Law enforcement, Counterterrorism, and Anti-Trafficking – Presents tools from those designing the computing and engineering techniques, architecture or policies related to applications confronting radicalisation, terrorism, and trafficking.

The books pertain to engineers working in law enforcement and researchers who are researching on capabilities of LEAs, though the series is truly multidisciplinary – each book will have hard core computer science, application of ICT in security and security / policing domain chapters. The books strike a balance between theory and practice.

Iman Almomani • Leandros A. Maglaras •
Mohamed Amine Ferrag • Nick Ayres
Editors

Cyber Malware

Offensive and Defensive Systems

Springer

Editors
Iman Almomani
Security Engineering Lab
Prince Sultan University
Riyadh, Saudi Arabia

Computer Science Department
The University of Jordan
Amman, Jordan

Mohamed Amine Ferrag
AI and Digital Science Research Center
Technology Innovation Institute
Masdar City, Abu Dhabi, United Arab Emirates

Leandros A. Maglaras
School of Computing
Edinburgh Napier University
Edinburgh, UK

Nick Ayres
School of Computer Science and Informatics
De Montfort University
Leicester, UK

ISSN 2523-8507 ISSN 2523-8515 (electronic)
Security Informatics and Law Enforcement
ISBN 978-3-031-34968-3 ISBN 978-3-031-34969-0 (eBook)
https://doi.org/10.1007/978-3-031-34969-0

This Springer imprint is published by the registered company Springer Nature Switzerland AG
The registered company address is: Gewerbestrasse 11, 6330 Cham, Switzerland

Paper in this product is recyclable.

Preface

The threat landscape is changing very quickly. With billions of connected IoT devices, mostly reactive detection and mitigation strategies, and finally big data challenges, we face an extremely rapidly expanding attack surface with a variety of attack vectors, a clear asymmetry between attackers and defenders, and a rapidly expanding attack surface. Additional arguments suggest that cybersecurity approaches must be rethought in terms of reducing the attack surface, making the attack surface dynamic, automating detection, risk assessment, and mitigation, and investigating the prediction and prevention of malware attacks with the use of emerging technologies like blockchain, artificial intelligence, and machine learning. Additionally, there is a clear asymmetry of attacks and an enormous amount of data.

This book provides the foundational aspects of malware attack vectors and appropriate defense mechanisms against malware. In addition, the book equips you with the necessary knowledge and techniques to successfully lower risk against emergent malware attacks. The book discusses both theoretical, technical, and practical issues related to malware attacks and defense making it an ideal reading material.

Many aspects motivated the decision toward the creation of this book. As mentioned in recent threat landscape reports, malware is on the rise again after the decrease that was noticed and linked to COVID-19. Malicious actors frequently employ malware in their campaigns. Gaining and maintaining control of assets, evading and deceiving defenses, and carrying out post-compromise actions all require this fundamental capability. The book has two clear goals. The first is to bring in front important security problems that arise in the advent of malware, and the second is to highlight a variety of possible solution approaches that might be able to address them. Specialists and experts present their significant efforts to fulfill these goals.

This book starts with an introductory chapter about the emerging trends of cyber-malware, and then it includes nine chapters that are organized into the following three parts:

Part 1 presents solutions on Android OS malware static features extraction and detection of Android malware applications.
Part 2 contains many applications that use artificial intelligence for detecting fast flux service networks and malware.
Part 3 presents and discusses techniques that can be used for IoT and cloud malware analysis.

In Chap. 1, an effective vision-based multi-classification system for detecting various malware families in Android apps is presented. Malware in Android apps could be detected using the proposed system in visual color or grayscale formats. The tested evaluation metrics and acquired detection results performed in the chapter demonstrate that the proposed vision-based system is a promising option for Android OS malware analysis.

Based on network traffic behavior analysis, Chapter 2 proposes a novel privacy-preserving federated deep learning method that makes use of convolutional neural networks (CNN) to identify various kinds of malware. The proposed detection method is evaluated in terms of detection rate, accuracy, and performance under various federated learning settings.

The third version of the Android automatic Static Parsing tool (ASParse-V3), and its integration with other detection methods are discussed in Chap. 3 in terms of the significance of static analysis for feature extraction, dataset generation, and malware analysis systems. The results of the analysis can be fed to deep learning models and machine learning algorithms for malware analysis and detection. In addition, Android OS applications were used to demonstrate the system's capabilities.

The fast flux architecture, operation, and characterization of FFSNs are the primary topics of discussion in Chap. 4. In addition, the chapter provides a summary of fast flux detection mechanisms, highlighting the most significant difficulties and potential future research directions.

A static, graph-based approach is presented in Chap. 5 that uses machine learning to classify executable samples into malicious or benign API Call Graphs. A measure of the Abstract API Call Graph's similarity to the samples of a given dataset, which include labeled samples of malware and benign samples, is calculated by the proposed method. Additionally, it divides the similarity vector space and performs classification using the support vector machine (SVM) algorithm. Both unweighted and weighted Abstract API Call Graphs are used to evaluate the method, demonstrating high accuracy.

Chapter 6 gives a thorough survey of cutting-edge deep learning-based malware analysis and detection solutions focusing on Microsoft Windows, over the time of 2015–2022. The section gives a detailed scientific classification that classifies these solutions as per different measures including the investigation task, the analysis task, the nature of the extracted features, the used features representation method, and the used deep learning algorithms. Besides, the section talks about these solutions concerning the size and nature of the testing dataset, the performance evaluation metrics for the various tasks, and the accomplished outcomes.

Threats to the Internet of Things (IoT) and smart systems are covered in Chap. 7, as is a brief overview of malware detection and evasion techniques. For the IoT and smart systems to be utilized to their full potential, it is essential to investigate novel cyberattacks while simultaneously developing and implementing countermeasures. The objectives of this chapter are to investigate various strategies for the detection and evasion of cybersecurity threats in the IoT domain as well as evaluate security issues that are anticipated to limit IoT deployment.

In Chap. 8, a method for multiclass classification employing XG-Boost and CatBoost to classify the intrusion attack's category type is proposed. The proposed strategy aimed to develop a recent multiclass classification to classify the category type labels of IoT intrusion attacks. Precision, recall, f1-score, and G-mean were used to evaluate the experiments, which were then compared to other basic classifiers.

Malware attacks and methods for preventing malware threats in cloud computing architecture are examined in Chap. 9. Data breaches, malicious insiders, man-in-the-middle attacks, denial-of-service (DOS) and distributed denial-of-service (DDOS) attacks, cookie poisoning attacks, and wrapping attacks are among the most frequently reported security threats, according to the study. The majority of these attacks are the result of multiple malware variants.

Riyadh, Saudi Arabia — Iman AlMomani
Edinburgh, UK — Leandros A. Maglaras
Masdar City, Abu Dhabi, United Arab Emirates — Mohamed Amine Ferrag
Leicester, UK — Nick Ayres

Introduction: Emerging Trends in Cyber-Malware

Cyber-malware refers to malicious software that is designed to damage or gain unauthorized access to computer systems, networks, and data. Cyber-malware has become a significant threat to individuals, businesses, and governments worldwide, and its impact can be devastating [1].

The history of cyber-malware dates back to the 1970s when the first computer virus, known as the Creeper virus, was created as an experimental program. The Creeper virus was designed to move between computers on a network and display the message "I'm the Creeper, catch me if you can." The first antivirus software, known as the Reaper, was then created to remove the Creeper virus from infected computers [2].

In the 1980s, as personal computers became more popular, cybercriminals began developing malware to exploit vulnerabilities in operating systems and software. In 1986, the first computer worm, known as the Morris worm, was created by a graduate student named Robert Morris. The Morris worm caused widespread damage to computer systems and resulted in significant financial losses. This incident prompted the creation of the Computer Emergency Response Team (CERT), which provides guidance and support for organizations affected by cyber-attacks [3].

In the 1990s, cybercriminals began developing more sophisticated malware, such as Trojans and keyloggers, to steal sensitive information from individuals and businesses [4]. The first known ransomware attack, known as the AIDS Trojan, was also created in 1990. The AIDS Trojan would encrypt the victim's files and demand payment in exchange for the decryption key.

In the 2000s, cyber-malware attacks became more prevalent, with high-profile incidents such as the ILOVEYOU virus and the Code Red worm causing significant damage to computer systems worldwide [5]. Additionally, cybercriminals began using social engineering techniques, such as phishing emails and fake websites, to trick individuals into giving up their login credentials and other sensitive information.

One of the most significant cyber-attacks involving malware in recent years was the SolarWinds attack. In December 2020, it was discovered that Russian hackers had gained access to the computer systems of several US government agencies and private companies by exploiting a vulnerability in the SolarWinds software [6]. The malware used in the attack, known as Sunburst, was a sophisticated piece of software that allowed the hackers to access sensitive information and carry out other malicious activities undetected for months.

Another recent cyber-attack involving malware was the Colonial Pipeline hack. In May 2021, a group of cybercriminals known as DarkSide used ransomware to gain access to the computer systems of Colonial Pipeline, a major US fuel pipeline operator. The attack forced the company to shut down its pipeline, causing widespread fuel shortages and price hikes across the eastern United States. The group demanded a ransom of $4.4 million in Bitcoin, which Colonial Pipeline ultimately paid [7].

In March 2021, Microsoft announced that Chinese hackers had been using malware to target organizations around the world. The hackers were exploiting four zero-day vulnerabilities in Microsoft Exchange Server, a popular email and collaboration platform used by many businesses and organizations [8]. The hackers used the malware to steal data and carry out other malicious activities, and the attack affected thousands of organizations in at least 115 countries.

In April 2021, cybersecurity researchers discovered a new type of malware known as Silver Sparrow. Unlike many other types of malware, Silver Sparrow was designed to target Apple computers, and it was found on nearly 30,000 Macs around the world [9]. While the malware was not actively causing any harm, its presence on so many devices was a cause for concern.

In recent years, cybercriminals have continued to evolve their tactics, with the development of more sophisticated ransomware, such as the WannaCry and NotPetya attacks, and the rise of cryptojacking, which involves using the victim's computer to mine cryptocurrency without their knowledge or consent [10].

As technology continues to advance, cyber-malware attacks are likely to become even more sophisticated and difficult to detect. However, cybersecurity professionals and organizations are also developing new tools and strategies to combat cyber-malware and protect against future attacks. So, the evolution of cyber-malware has been marked by increasingly sophisticated attacks and techniques. From the early days of the Creeper virus to the modern-day threats of ransomware and cryptojacking, cybercriminals have continuously adapted their tactics to exploit vulnerabilities in computer systems and networks [11]. However, through collaboration and innovation in cybersecurity, individuals, businesses, and governments can work to stay one step ahead of cyber-malware threats.

Individuals can become victims of cyber-malware through various means, including phishing emails, infected downloads, and social engineering attacks. Once the malware infects an individual's device, it can steal sensitive information such as login credentials, financial information, and personal data [3–5]. In some cases, cyber-malware can lock users out of their devices and demand payment for the return of access, also known as ransomware.

Businesses are at an even higher risk of cyber-malware attacks, as they often store large amounts of sensitive data that can be targeted by cybercriminals. The impact of cyber-malware on businesses can range from financial losses to reputational damage [7]. For instance, if a company's financial data is breached, it can result in significant financial losses and a loss of customer trust. Additionally, if a company's reputation is damaged due to a cyber-attack, it can lead to a decline in sales and revenue.

Governments are also vulnerable to cyber-malware attacks, as they often store classified information and sensitive data. A cyber-attack on a government's system can have severe consequences, including the theft of sensitive information, disruption of essential services, and even sabotage [9]. In some cases, cyber-malware attacks on governments have been carried out by state-sponsored hackers, leading to tensions between nations.

The impact of cyber-malware on individuals, businesses, and governments is not limited to financial losses and reputational damage. Cyber-malware attacks can also result in a loss of privacy, psychological distress, and physical harm. For instance, cyber-malware can be used to gain access to medical devices and cause harm to patients or to disrupt critical infrastructure and cause widespread power outages [8–11].

To protect against cyber-malware, individuals, businesses, and governments must take proactive measures to secure their systems and data.

This includes implementing strong passwords, keeping software up-to-date, using anti-virus software, and educating employees and users about cyber threats. Additionally, governments must work together to develop international frameworks and regulations to combat cyber-malware and hold cybercriminals accountable for their actions. Thus, cyber-malware is a growing threat to individuals, businesses, and governments worldwide [12]. The impact of cyber-malware can range from financial losses to reputational damage and can even result in physical harm. To protect against cyber-malware, it is essential to take proactive measures to secure systems and data and to work together to develop international frameworks and regulations to combat cybercrime.

Cyber-malware attacks can target a wide range of individuals, businesses, and organizations. However, certain targets are more commonly targeted by cybercriminals due to their vulnerability or potential for financial gain. Some of the common targets of cyber-malware include [13]:

- *Individuals:* Cybercriminals often target individuals with phishing emails or malware disguised as legitimate software. Individuals can be targeted for their personal information, such as login credentials, banking information, and social security numbers. Additionally, cybercriminals may use malware to gain access to an individual's computer system, allowing them to steal sensitive information or use the victim's computer for illegal activities.
- *Small businesses:* Small businesses are often targeted by cybercriminals due to their limited resources and lack of robust cybersecurity measures. Small businesses may be targeted for their financial information, customer data, or intellectual property. Ransomware attacks are also common among small businesses, as cybercriminals may demand payment in exchange for restoring access to the victim's files or computer system.
- *Large corporations:* Large corporations are also common targets of cyber-malware attacks, as they may hold valuable intellectual property or financial information. Cybercriminals may use malware to gain unauthorized access to a corporation's network or use phishing emails to trick employees into giving up sensitive information.
- *Government agencies:* Government agencies are often targeted by cybercriminals seeking sensitive information or attempting to disrupt government operations. Cyber-malware attacks on government agen-

cies can result in the theft of classified information, disruption of critical infrastructure, and other significant consequences.
- *Healthcare providers:* Healthcare providers are another common target of cyber-malware attacks, as they may hold sensitive patient information, including medical records and billing information. Cybercriminals may use malware to gain unauthorized access to a healthcare provider's network or steal patient data for identity theft or insurance fraud.

Consequently, staying up to date with new trends in cyber-malware is incredibly important in today's digital age. With the increasing number of devices and networks connected to the internet, the threat of cyber-attacks is more prevalent than ever before. Malware, short for malicious software, is designed to damage, disrupt, or gain unauthorized access to computer systems. The technology used by cybercriminals is continually evolving, and new types of malware are being developed all the time. By staying up to date with the latest trends in cyber-malware, you can ensure that you are better prepared to defend against attacks and protect your digital assets.

One of the most significant reasons to stay up to date with cyber-malware trends is to identify new threats before they become widespread. Cybercriminals often use new malware to exploit vulnerabilities in systems before antivirus software and other security measures can be updated to address the threat [14]. By being aware of new types of malware, you can take steps to protect yourself and your organization before an attack occurs.

Another reason to stay up to date with cyber-malware trends is to keep your security measures current. As new malware is developed, antivirus software and other security measures are updated to protect against them. By staying informed about new threats, you can ensure that your security measures are up-to-date and effective [15]. Failure to update your security measures can leave your devices and networks vulnerable to attack.

Additionally, staying up to date with cyber-malware trends can help you stay ahead of the competition. Cybersecurity is becoming increasingly important in today's digital landscape, and companies that fail to take it seriously may suffer reputational damage or lose customers. By demonstrating that you are aware of the latest threats and taking steps to protect your digital assets, you can build trust with your customers and gain a competitive advantage.

In conclusion, staying up to date with new trends in cyber-malware is essential to protect yourself, your organization, and your customers from cyber-attacks. By being aware of new threats and keeping your security measures current, you can stay one step ahead of cybercriminals and avoid the potentially devastating consequences of a successful cyber-attack.

Malware Analysis Techniques

Cyber-malware, also known as malicious software, is a type of software designed to infiltrate, damage, or disrupt computer systems, networks, and devices. Cyber-malware is often used for criminal purposes, such as stealing sensitive information or extorting money from victims.

Common Types of Cyber-Malware

There are several types of cyber-malware, including [13, 14]:

- *Virus:* A computer virus is a type of malware that infects a computer system by inserting its code into legitimate programs or documents. Once infected, the virus can replicate itself and spread to other systems, causing damage and stealing sensitive information.
- *Trojan:* A Trojan is a type of malware that disguises itself as legitimate software, often through email attachments or downloads. Once installed, the Trojan can allow cybercriminals to gain unauthorized access to the victim's computer, steal sensitive data, and even take control of the system.
- *Worm:* A worm is a self-replicating malware that spreads through networks and can cause significant damage to computer systems and networks. Worms often exploit vulnerabilities in software or operating systems, allowing cybercriminals to gain unauthorized access and steal sensitive information.
- *Ransomware:* Ransomware is a type of malware that encrypts the victim's files or computer system, rendering it unusable. The cybercriminals then demand payment, often in cryptocurrency, to provide the decryption key and restore access to the victim's data or system.
- *Adware:* Adware is a type of malware that displays unwanted or intrusive advertisements on the victim's computer system. Adware can also collect personal information, browsing history, and search queries for targeted advertising purposes.
- *Spyware:* Spyware is a type of malware that collects sensitive information, such as login credentials, browsing history, and personal data, without the victim's knowledge or consent. Cybercriminals can

then use this information for identity theft, financial fraud, or other malicious purposes.

- *Rootkit:* A rootkit is a type of malware that allows cybercriminals to gain administrative access to the victim's computer system. Rootkits often remain hidden from antivirus software and can be difficult to detect and remove, allowing cybercriminals to maintain access to the victim's system for an extended period.

To conclude, cyber-malware is a type of malicious software that can cause significant damage and disrupt computer systems, networks, and devices. There are several types of cyber-malware, including viruses, Trojans, worms, ransomware, adware, spyware, and rootkits [16, 17]. Understanding the different types of cyber-malware and taking proactive measures to protect against them is essential for individuals, businesses, and governments.

Dynamic and Static Analysis

Dynamic and static analysis are two techniques commonly used in cybersecurity to detect and analyze malware [17]. Static analysis involves examining the code of a program or file without actually executing it. This can involve using specialized tools and techniques to scan the code for known patterns or characteristics of malware. Static analysis is often used as a first step in malware analysis to quickly identify potential threats and determine whether further analysis is necessary.

Dynamic analysis, on the other hand, involves executing the program or file in a controlled environment to observe its behavior. This can involve running the program or file in a virtual machine or sandboxed environment to prevent any harm to the host system. Dynamic analysis can provide more detailed information on the behavior of malware, including its interactions with the operating system, network connections, and other processes.

Both dynamic and static analysis have their advantages and limitations [18]. Static analysis is often faster and less resource-intensive than dynamic analysis, making it a useful tool for quickly identifying potential threats. However, static analysis may not always be able to detect more advanced or sophisticated malware that is designed to evade detection.

Dynamic analysis, on the other hand, provides a more comprehensive view of the behavior of malware, which can be useful in understanding how the malware operates and identifying potential vulnerabilities in the system.

However, dynamic analysis can be more time-consuming and resource-intensive than static analysis, as it requires the execution of the malware in a controlled environment.

In practice, both dynamic and static analysis are often used in combination to provide a more comprehensive view of malware and its behavior. By using both techniques, cybersecurity professionals can quickly identify potential threats using static analysis, and then perform more detailed analysis using dynamic analysis to gain a deeper understanding of the malware's behavior and potential impact on the system [17, 18].

Malware Debugging Techniques

Malware authors are constantly evolving their techniques to evade detection and infect systems, which means that malware analysts need to constantly develop new techniques to detect and remove malware. One such technique is malware debugging [19].

Malware debugging is the process of analyzing malware by examining its code in a controlled environment. This allows analysts to identify the malware's behavior, the techniques it uses to evade detection, and the vulnerabilities it exploits. Several techniques can be used in malware debugging, including [20]:

- *Disassembly:* Disassembling the malware code is the process of converting the binary executable code into human-readable assembly code. This technique can help malware analysts to understand the behavior of the malware and identify potential vulnerabilities that it exploits.
- *Debugging tools:* Debugging tools, such as OllyDbg, IDA Pro, and WinDbg, can be used to analyze malware by allowing analysts to step through the code, set breakpoints, and view the contents of memory and registers. These tools can help to identify how the malware communicates with its command and control server, the files it creates on the infected system, and other behaviors that it exhibits.
- *Virtual machines:* Malware can be run in a virtual machine environment, such as VirtualBox or VMWare, to create a controlled environment for analysis. This technique can help to isolate the malware from the rest of the system, preventing it from infecting other files and processes on the host machine.

- *Sandboxing:* A sandbox is a virtual environment that isolates the malware from the rest of the system. This technique can help to prevent the malware from infecting other files and processes on the host machine while still allowing malware analysts to observe its behavior.
- *Dynamic analysis:* Dynamic analysis involves observing the malware as it runs in a controlled environment. This technique can help to identify the malware's behavior, such as the files it creates, the registry keys it modifies, and the network connections it establishes.
- *Code injection:* Code injection involves injecting code into the malware's process to modify its behavior or to observe its interactions with the operating system. This technique can help to identify the malware's communication with its command and control server, the data it exfiltrates, and other behaviors that it exhibits.

Identifying Malware Behavior

Identifying malware behavior is a critical step in malware analysis, as it can help security professionals to understand how a malware infection works and develop strategies for mitigating its impact. Malware behavior can include a range of activities, such as modifying system settings, stealing data, and communicating with remote servers. Here are some common techniques used to identify malware behavior [21–23]:

- *Static analysis:* This involves examining the malware code without actually running it. This can be done by examining the binary file or the source code and can help to identify the malware's behavior by looking at functions and routines used by the malware. Static analysis can also be used to identify specific strings or signatures associated with the malware.
- *Dynamic analysis:* This involves running the malware in a controlled environment to observe its behavior. This can be done in a sandbox, virtual machine, or other isolated environment. Dynamic analysis can help to identify the malware's activities, such as files it creates, registry keys it modifies, network connections it makes, and commands it sends or receives.
- *Network traffic analysis:* This involves monitoring network traffic to identify unusual activity. This can include unusual data transfers, unusual ports or protocols, and unusual server activity. Network traffic

analysis can help to identify malware that is communicating with remote servers.
- *Endpoint detection and response (EDR):* EDR tools monitor activity on endpoints (such as desktops, servers, and mobile devices) to detect suspicious behavior. EDR tools can identify indicators of compromise (IoCs), such as suspicious processes, changes to the registry or file system, and attempts to bypass security controls.
- *Reverse engineering:* This involves decompiling or disassembling the malware code to identify its behavior. Reverse engineering can help to identify how the malware communicates with its command and control server, how it encrypts or decrypts data, and how it modifies system settings.
- *Memory analysis:* This involves examining the contents of the computer's memory to identify malware behavior. Memory analysis can help to identify malware that has been loaded into memory and identify any unusual processes or network connections.
- *Behavioral analysis:* This involves observing the malware's behavior in a virtual environment to identify any unusual or malicious activity. Behavioral analysis can help to identify the specific behavior of the malware, which can be used to develop targeted mitigation strategies.

In conclusion, identifying malware behavior is an important step in malware analysis. It involves using a combination of techniques, such as static analysis, dynamic analysis, network traffic analysis, endpoint detection and response, reverse engineering, memory analysis, and behavioral analysis, to identify the malware's activities and develop strategies for mitigating its impact [24]. By understanding the behavior of malware, security professionals can better protect their systems and networks against malware infections.

Malware Distribution Methods

Malware distribution methods refer to the various ways in which malicious software is disseminated to infect systems and devices. Malware can take many forms, including viruses, worms, Trojans, ransomware, and spyware, among others. Malware authors often use multiple distribution methods to increase the likelihood of infecting as many devices as possible. The most common malware distribution methods are as follows [25, 26]:

- *Email Attachments:* Malware authors use emails to distribute malware by attaching malicious files to emails. The recipient is tricked into downloading and opening the attachment, which infects their system. The attachments may appear as legitimate files, such as a PDF or a Word document, but once opened, the malware executes on the system.
- *Social Engineering:* Social engineering involves tricking users into downloading or installing malware by using psychological manipulation techniques. For example, attackers may use a fake website to convince users to download an application that is malware. Social engineering may also involve using fake antivirus alerts or fake software updates to trick users into installing malware.
- *Drive-By Downloads:* Drive-by downloads involve malware being installed on a user's computer without their knowledge or consent when they visit a website. This is typically accomplished by exploiting vulnerabilities in the user's web browser or other software.
- *Malvertising:* Malvertising is the distribution of malware through online advertisements. Attackers use legitimate-looking advertisements to lure users into clicking on them, which then leads to the installation of malware on the user's computer.
- *Infected Software:* Malware authors sometimes distribute infected software or applications that appear legitimate but are infected with malware. Once the software is downloaded and installed, the malware executes on the system.
- *USB Drives:* Malware can also be distributed through USB drives that are infected with malware. When the USB drive is inserted into a computer, the malware automatically executes on the system.
- *Watering Hole Attacks:* In a watering hole attack, attackers infect a website that is frequently visited by their target audience. The attackers then wait for their targets to visit the infected website, where they are infected with malware.
- *Phishing:* This method involves sending emails or messages that appear to be from a trusted source but contain links to malicious websites or attachments that download malware onto the victim's computer.
- *Software vulnerabilities:* Cybercriminals can exploit vulnerabilities in legitimate software applications to install malware onto a victim's computer.

- *Malicious websites:* Cybercriminals can create malicious websites that contain malware. These websites may look legitimate, but they are designed to infect visitors' computers with malware. In some cases, simply visiting the website is enough to download the malware.
- *Social media:* Cybercriminals can use social media platforms to distribute malware. They may create fake profiles or pages that appear to be legitimate but contain links to infected websites or downloads.
- *File sharing networks:* Some malware is distributed through peer-to-peer (P2P) file-sharing networks. Cybercriminals may upload infected files, such as movies or music, and entice users to download them.
- *Mobile devices:* Malware can also be distributed through mobile devices, such as smartphones and tablets. Cybercriminals may create fake apps that contain malware or send infected links through text messages or social media.

Thus, malware distribution methods are constantly evolving, and attackers are becoming more sophisticated in their techniques. It is essential to remain vigilant when opening emails, downloading software, or visiting websites to avoid falling victim to malware. Keep your software updated, use reputable antivirus software, and be cautious of suspicious emails and websites. So, to protect against these malware distribution methods, it's important to keep software up to date, use antivirus software, be cautious when opening email attachments or clicking on links, and avoid downloading software from untrusted sources.

Malware Prevention and Mitigation Strategies

Malware prevention and mitigation strategies are essential in today's digital age, where malware threats are prevalent and continue to evolve. Prevention and mitigation strategies are measures put in place to reduce the likelihood and severity of potential hazards, disasters, or crises. These strategies aim to prevent or mitigate the negative impact of these events on individuals, communities, and the environment [27].

Prevention strategies involve taking measures to prevent an event from occurring. These strategies can include implementing safety measures, such as using protective equipment, conducting safety training, or installing safety features in buildings or equipment. Preventive strategies can also involve enforcing regulations or laws to deter risky behaviors or practices.

Mitigation strategies involve taking steps to reduce the impact of an event that has already occurred. These strategies can include emergency response plans, such as evacuation plans, first aid procedures, and disaster relief efforts. Mitigation strategies can also involve restoration efforts, such as rebuilding infrastructure or rehabilitating natural habitats [28].

Effective prevention and mitigation strategies are essential for reducing the impact of disasters and crises. By taking proactive steps to prevent events from occurring or mitigating their effects, we can reduce the risk of harm and save lives [29]. Additionally, these strategies can also help reduce the economic and environmental impact of disasters, making recovery and restoration efforts more manageable.

Examples of prevention and mitigation strategies include [27–29]:

- *Hazard assessments:* Conduct regular assessments to identify potential hazards and develop appropriate prevention and mitigation strategies.
- *Early warning systems:* Implement systems that provide early warning of potential hazards, such as natural disasters or industrial accidents, to allow for timely response and mitigation.
- *Infrastructure improvement:* Upgrade and maintain infrastructure, such as roads, bridges, and buildings, to make them more resilient to disasters.
- *Community education and outreach:* Educate communities about potential hazards, how to prepare for disasters, and what to do in case of emergency.
- *Disaster response planning:* Develop comprehensive plans for responding to disasters and crises, including evacuation plans, emergency communication systems, and disaster relief efforts.
- *Environmental protection measures:* Implement measures to protect the environment, such as reducing pollution and conserving natural resources, to prevent or mitigate the impact of disasters.
- *Risk assessments:* Conduct regular assessments to identify potential hazards and develop appropriate prevention and mitigation strategies.
- *Use Antivirus Software:* Install and regularly update a reputable antivirus software program on your computer or device. Antivirus software can help detect and remove malware from your system.
- *Keep Software Up-to-date:* Keep your operating system, web browser, and other software applications up-to-date with the latest security patches and updates. Cybercriminals often exploit vulnerabilities in outdated software.
- *Use Strong Passwords:* Use strong, unique passwords for all your accounts and avoid using the same password across multiple accounts.

Consider using a password manager to generate and store complex passwords.

- *Be Cautious of Email Attachments:* Do not open email attachments or click on links from unknown or suspicious sources. Malware can be spread through email attachments and links.
- *Enable Two-Factor Authentication:* Enable two-factor authentication (2FA) whenever possible. This adds an extra layer of security to your accounts by requiring a second form of authentication, such as a code sent to your mobile device.
- *Back Up Your Data:* Regularly back up your data to an external hard drive or cloud storage service. This can help mitigate the impact of malware if your system is infected.
- *Educate Yourself:* Stay informed about the latest threats and best practices for preventing and mitigating malware. Learn how to recognize and avoid phishing scams, and be cautious when downloading and installing software from the Internet.

Future of Cyber-Malware

The future of cyber-malware is a topic of concern for cybersecurity professionals and businesses worldwide. As technology continues to evolve and become more complex, so do the threats posed by cyber-malware. One trend that is likely to continue in the future is the use of artificial intelligence (AI) by cybercriminals to develop more sophisticated and effective malware. AI-powered malware can adapt to its environment, evade detection, and target specific vulnerabilities in a network or system. This type of malware can also learn from its actions and adjust its behavior accordingly, making it more difficult to stop.

Another potential development in cyber-malware is the increased use of ransomware attacks. Ransomware is a type of malware that encrypts a victim's files or data and demands payment in exchange for the decryption key. This type of attack has become increasingly common in recent years and is likely to continue in the future, as it can be highly profitable for attackers. In fact, some experts predict that ransomware attacks may become more targeted, with attackers focusing on specific industries or organizations with high-value data.

The Internet of Things (IoT) is another area of concern when it comes to the future of cyber-malware. IoT devices are often connected to the internet and can be vulnerable to attacks, as they may not have strong security protocols in place. As the number of IoT devices continues to grow, so does the potential for cyber-attacks targeting them. This could lead to large-scale disruptions, such as attacks on critical infrastructure or widespread data breaches.

Finally, there is a growing concern about the use of nation-state-sponsored cyber-malware attacks. Governments may use cyber-malware to gain access to sensitive information, disrupt rival countries' infrastructure, or carry out espionage activities. These attacks can be difficult to trace and may have significant political and economic consequences.

So, the future of cyber-malware is likely to be characterized by increasingly sophisticated and targeted attacks. As technology continues to advance, so do the threats posed by cybercriminals. To mitigate these risks, businesses and individuals must remain vigilant and take steps to protect their networks, devices, and data. This includes implementing strong security protocols, keeping software up-to-date, and educating users about the risks of cyber-malware.

Trends and Predictions for Future Malware Development

Malware, or malicious software, has been a persistent threat to computer systems and networks since the dawn of the internet. Cybercriminals constantly seek out new ways to exploit vulnerabilities in software and hardware to gain unauthorized access to sensitive data or control systems for nefarious purposes. In recent years, malware development has become more sophisticated, and new trends are emerging that could shape the future of cybercrime [30]. Here are some predictions for trends in malware development in the near future [31].

- *Fileless malware:* Fileless malware attacks are on the rise, and this trend is likely to continue in the coming years. Fileless malware, also known as memory-resident malware, operates entirely in a computer's memory and leaves no trace on the system's hard drive. This makes it difficult to detect and remove, as traditional antivirus software relies on scanning files on a hard drive. As more businesses adopt cloud-based computing and mobile devices become more prevalent, fileless malware is likely to become a more significant threat.
- *Malware as a service:* Malware as a service (MaaS) is a growing trend in the cybercriminal underground. Just like software as a service (SaaS), MaaS allows cybercriminals to rent or purchase malware

from a third-party provider. This lowers the barrier to entry for less technically savvy criminals, who can now launch sophisticated attacks without having to develop their own malware. As MaaS becomes more prevalent, we can expect to see more varied and sophisticated malware being developed and deployed.

- *Advanced evasion techniques:* As cybersecurity defenses become more sophisticated, malware developers are turning to advanced evasion techniques to avoid detection. These techniques include using encryption to hide malicious code, exploiting vulnerabilities in antivirus software, and creating polymorphic malware that can change its code to evade detection. As evasion techniques become more sophisticated, it will become increasingly difficult to detect and prevent malware attacks.
- *Targeted attacks:* Rather than launching mass attacks, cybercriminals are increasingly targeting specific individuals or organizations. This allows them to conduct more sophisticated attacks, such as spear-phishing, that are tailored to the victim's interests or behaviors. As more data becomes available on individuals and organizations, we can expect to see more targeted attacks that leverage this information to bypass defenses and gain access to sensitive data.
- *IoT malware:* With the rise of the Internet of Things (IoT), there is a growing concern about the security of these devices. IoT devices are often not designed with security in mind and can be easily hacked, giving cybercriminals access to sensitive data or control over critical infrastructure. As the number of IoT devices continues to grow, we can expect to see more malware specifically designed to target these devices.
- *Machine Learning-Based Malware:* Machine learning has become a powerful tool for cybersecurity, and malware developers are no exception. By using machine learning algorithms, malware can adapt to its environment and learn how to evade detection.
- *Deepfakes:* Deepfakes are videos or images that have been manipulated using artificial intelligence to make them appear real. In the future, we can expect to see more malware that uses deepfakes to trick users into downloading or installing malicious software.
- *Mobile Malware:* With the increasing use of mobile devices, mobile malware has become a growing concern. In the future, we can expect to see more mobile-specific malware that can steal sensitive data or take control of the device.

Consequently, malware development is constantly evolving, and new trends and techniques are emerging all the time. As cybersecurity defenses become more sophisticated, cybercriminals will continue to find new ways to bypass them. Individuals and organizations need to stay informed about the latest trends in malware development and take appropriate measures to protect their systems and data.

Emerging Threats and Attack Vectors

As the digital landscape continues to evolve, so do the threats and attack vectors that cybercriminals use to compromise systems and steal data. Here are some emerging threats and attack vectors to be aware of [30, 31]:

- *Supply Chain Attacks:* Supply chain attacks involve targeting a third-party vendor that supplies software or hardware components to a larger organization. The attackers compromise the vendor's systems, injecting malware into the products or services that the vendor provides. When the larger organization installs or uses the compromised product or service, the malware spreads to their systems, giving the attackers access to sensitive data.
- *Zero-Day Exploits:* Zero-day exploits are vulnerabilities in software or hardware that are unknown to the vendor or manufacturer. Attackers exploit these vulnerabilities before the vendor can patch them, giving them access to the affected systems. Zero-day exploits are particularly dangerous because there are no known defenses against them.
- *Phishing:* Phishing attacks are social engineering attacks that attempt to trick users into revealing sensitive information or installing malware. Phishing attacks can take many forms, including emails, text messages, or phone calls. These attacks are becoming increasingly sophisticated, using tactics such as personalized messaging and spoofing trusted sources.
- *Ransomware:* This is a type of malware that encrypts an organization's data and demands payment in exchange for the decryption key. Ransomware attacks are becoming increasingly common and can cause significant disruption and financial losses.
- *Social engineering:* Social engineering attacks involve tricking individuals into divulging sensitive information or performing an action that compromises the security of an organization. Common techniques include phishing emails and pretexting.

- *Machine learning attacks:* Machine learning algorithms are vulnerable to attack, which can lead to inaccurate predictions or even malicious behavior. Adversarial attacks, where an attacker deliberately modifies data to trick the algorithm, are becoming increasingly common.
- *Insider threats:* Insider threats can be intentional or unintentional, but they can cause significant damage to an organization's security. Organizations need to implement policies and procedures to detect and prevent insider threats.
- *AI-Powered Attacks:* As artificial intelligence (AI) becomes more prevalent in cybersecurity, attackers are using AI-powered tools to automate attacks. AI can be used to automate phishing attacks, identify vulnerabilities, and evade detection by security measures.
- *Cloud-Based Attacks:* Cloud computing has become a popular choice for businesses, but it has also created new attack vectors for cyber-criminals. Cloud-based attacks can include exploiting vulnerabilities in cloud infrastructure, stealing login credentials, or compromising data stored in the cloud.
- *Internet of Things (IoT) Attacks:* IoT devices, such as smart home devices and industrial control systems, are becoming more prevalent in our lives. However, these devices often have weak security measures and are vulnerable to attack. Attackers can use IoT devices to launch attacks, such as Distributed Denial of Service (DDoS) attacks, or to steal sensitive data.

Therefore, as technology continues to advance, cybercriminals will continue to find new and more sophisticated ways to compromise systems and steal data. It is important to stay informed about emerging threats and attack vectors and take proactive measures to protect our systems and data. This includes regularly updating software, using strong passwords, and implementing multi-factor authentication.

The Role of Artificial Intelligence in Malware Development and Detection

Artificial intelligence (AI) is playing an increasingly important role in both malware development and detection. On the one hand, AI can be used to create more sophisticated and effective malware, while on the other hand, it can also be used to develop more advanced detection and prevention techniques [32].

One of how AI is being used in malware development is through the use of machine learning algorithms. By training these algorithms on large datasets of existing malware, researchers can develop new malware that is specifically designed to evade existing detection methods. Machine learning can also be used to create more sophisticated attack strategies, such as spear-phishing campaigns that are tailored to individual victims.

However, AI is also being used to develop new and more effective methods for detecting and preventing malware. For example, AI can be used to analyze network traffic and identify patterns of behavior that are indicative of a malware infection. Similarly, machine learning algorithms can be trained to identify specific features of malware code, making it possible to detect and block new malware strains as they emerge.

Another area where AI is having a significant impact on malware detection is in the development of so-called "next-generation" antivirus (NGAV) solutions. These solutions use a combination of machine learning algorithms and behavioral analysis techniques to detect and block malware in real time, even if it has never been seen before. NGAV solutions can also be used to identify and block previously unknown attack vectors, such as zero-day exploits, that traditional antivirus solutions are unable to detect [33].

Here are some of the new research trends for the role of AI in malware development and detection [32, 33]:

- *Adversarial machine learning:* Adversarial machine learning is a technique where an attacker deliberately modifies data to trick the machine learning algorithm into making a wrong prediction. In the context of malware detection, attackers can use this technique to evade detection by creating malware that appears benign to machine learning algorithms. New research is exploring how to develop machine learning algorithms that are more resilient to adversarial attacks.
- *Explainable AI:* Explainable AI is a technique that enables humans to understand how a machine learning algorithm is making its predictions. In the context of malware detection, explainable AI can help security analysts understand how a particular malware was detected and what features of the malware triggered the detection. This can help security analysts develop more effective detection strategies.
- *Deep learning:* Deep learning is a subfield of machine learning that involves training deep neural networks with multiple layers. New research is exploring how deep learning can be used to detect malware

by analyzing its behavior. For example, deep learning can be used to analyze network traffic and identify patterns of behavior that are indicative of a malware infection.

- *Reinforcement learning:* Reinforcement learning is a type of machine learning where an algorithm learns to make decisions by interacting with an environment. In the context of malware detection, reinforcement learning can be used to train an algorithm to make decisions about whether a particular file is malware or not based on feedback from the environment.
- *Generative adversarial networks (GANs):* GANs are a type of deep learning algorithm that consists of two neural networks that compete against each other. One network generates samples, while the other network tries to distinguish between real and fake samples. In the context of malware detection, GANs can be used to generate synthetic malware samples that can be used to train machine learning algorithms.
- *Transfer learning:* Transfer learning is a technique that involves training a machine learning algorithm on one task and then transferring that knowledge to another task. In the context of malware detection, transfer learning can be used to train a machine learning algorithm on a large dataset of non-malicious software and then transfer that knowledge to detect malware.

In conclusion, while AI is being used to create more sophisticated and effective malware, it is also playing an important role in the development of new and more advanced malware detection and prevention techniques. As the threat landscape continues to evolve, AI will likely play an increasingly important role in both offensive and defensive cybersecurity strategies.

Conclusions and Future Work

This chapter highlighted some of the latest trends and challenges in the field of malware detection and prevention. One of the key takeaways from this chapter is the increasing sophistication and complexity of malware attacks. Malware developers are becoming more adept at evading detection and are using more advanced techniques like artificial intelligence and machine learning to develop new strains of malware. As a result, traditional malware detection methods are becoming less effective. To combat this evolving threat landscape, researchers are exploring new approaches to malware detection and prevention. These include the use of artificial intelligence and machine learning algorithms to analyze network traffic and identify patterns of behavior that are indicative of a malware infection. Next-generation antivirus solutions that use a combination of machine learning and behavioral analysis techniques are also emerging as an important defense against malware attacks.

Another important trend highlighted in this chapter is the increasing importance of collaboration between industry, academia, and government in the fight against cyber-malware. By working together and sharing information, researchers and cybersecurity professionals can stay ahead of emerging threats and develop more effective countermeasures. Looking to the future, the chapter concludes by suggesting that the field of malware detection and prevention will continue to evolve rapidly. New techniques and approaches will be developed to combat increasingly sophisticated attacks, and the role of artificial intelligence and machine learning in this field will continue to grow. In addition, the rise of the IoT is expected to introduce new challenges for malware detection and prevention, as these devices often lack the security features of traditional computers and servers.

In conclusion, this chapter provided a valuable overview of the latest trends and challenges in the field of malware detection and prevention. By staying abreast of these trends and developing new and innovative solutions, cybersecurity professionals can help to protect individuals, businesses, and organizations against the growing threat of cyber-malware.

Security Engineering Lab, Computer Science Department
Prince Sultan University, Riyadh, Saudi Arabia
Electronics and Electrical Communication Engineering Department, Faculty of Electronic Engineering
Menoufia University, Menouf, Egypt
e-mail: welshafai@psu.edu.sa; walid.elshafai@el-eng.menofia.edu.eg

Walid El-Shafai

Security Engineering Lab,
Prince Sultan University, Riyadh,
Saudi Arabia
Computer Science Department,
The University of Jordan, Amman, Jordan
e-mail: imomani@psu.edu.sa;
i.momani@ju.edu.jo

Iman Almomani

School of Computing
Edinburgh Napier University, Edinburgh,
UK
e-mail: l.maglaras@napier.ac.uk

Leandros A. Maglaras

References

1. Aziz S, Irshad M, Haider SA, Wu J, Deng DN, Ahmad S (2022) Protection of a smart grid with the detection of cyber-malware attacks using efficient and novel machine learning models. Front Energy Res 10:1102
2. Choi KS, Lee CS, Merizalde J (2023) Spreading viruses and malicious codes. In: Handbook on crime and technology. Edward Elgar Publishing, Florida, United States, pp 232–250
3. Riebe T, Kaufhold MA, Reuter C (2021) The impact of organizational structure and technology use on collaborative practices in computer emergency response teams: an empirical study. Proc ACM Hum-Comput Interact 5(CSCW2):1–30
4. Gazet A (2010) Comparative analysis of various ransomware virii. J Comput Virol 6:77–90
5. Bridges L (2008) The changing face of malware. Netw Secur 2008(1):17–20
6. Alkhadra R, Abuzaid J, AlShammari M, Mohammad N (2021) Solar winds hack: in-depth analysis and countermeasures. In: 2021 12th international conference on computing communication and networking technologies (ICCCNT). IEEE, Kharagpur, India, pp 1–7

7. Danisevskis J, Piekarska M, Seifert JP (2014) Dark side of the shader: mobile gpu-aided malware delivery. In: Information security and cryptology–ICISC 2013: 16th international conference, Seoul, Korea, November 27–29, 2013, Revised Selected Papers 16. Springer International Publishing, Seoul, Korea, pp 483–495
8. Singh UK, Joshi C, Kanellopoulos D (2019) A framework for zero-day vulnerabilities detection and prioritization. J Inf Secur Appl 46:164–172
9. Kumar R, Bawa SR (1979) Hepatic silver binding protein (Ag BP) from sparrow (Passer domesticus). Experientia 35:1621–1623
10. Lika RA, Murugiah D, Brohi SN, Ramasamy D (2018) NotPetya: cyber attack prevention through awareness via gamification. In: 2018 International conference on smart computing and electronic enterprise (ICSCEE). IEEE, Shah Alam, Malaysia, pp 1–6
11. Warmsley D, Waagen A, Xu J, Liu Z, Tong H (2022) A survey of explainable graph neural networks for cyber malware analysis. In: 2022 IEEE international conference on big data (big data). IEEE, Osaka, Japan, pp 2932–2939
12. Eboibi FE (2017) A review of the legal and regulatory frameworks of Nigerian Cybercrimes Act 2015. Comput Law Secur Rev 33(5):700–717
13. Wu M, Moon YB (2017) Taxonomy of cross-domain attacks on cybermanufacturing system. Procedia Comput Sci 114:367–374
14. Patel P, Kannoorpatti K, Shanmugam B, Azam S, Yeo KC (2017) A theoretical review of social media usage by cyber-criminals. In: 2017 International conference on computer communication and informatics (ICCCI). IEEE, Coimbatore, India, pp 1–6
15. Rogers MK (2011) The psyche of cybercriminals: a psycho-social perspective. In: Cybercrimes: a multidisciplinary analysis, Springer, Singapore, pp 217–235
16. Almomani I, Ahmed M, El-Shafai W (2022) Android malware analysis in a nutshell. PloS One 17(7):e0270647
17. Almomani I, Alkhayer A, El-Shafai W (2022) An automated vision-based deep learning model for efficient detection of android malware attacks. IEEE Access 10:2700–2720
18. El-Shafai W, Almomani I, AlKhayer A (2021) Visualized malware multi-classification framework using fine-tuned CNN-based transfer learning models. Appl Sci 11(14):6446

19. Afianian A, Niksefat S, Sadeghiyan B, Baptiste D (2019) Malware dynamic analysis evasion techniques: a survey. ACM Comput Surv (CSUR) 52(6):1–28
20. Chen P, Huygens C, Desmet L, Joosen W (2016) Advanced or not? A comparative study of the use of anti-debugging and anti-VM techniques in generic and targeted malware. In: ICT systems security and privacy protection: 31st IFIP TC 11 international conference, SEC 2016, Ghent, Belgium, May 30–June 1, 2016, Proceedings 31. Springer International Publishing, Ghent, Belgium, pp 323–336
21. Yu B, Fang Y, Yang Q, Tang Y, Liu L (2018) A survey of malware behavior description and analysis. Front Inf Technol Electron Eng 19:583–603
22. Shaid SZM, Maarof MA (2014) Malware behavior image for malware variant identification. In: 2014 International symposium on biometrics and security technologies (ISBAST). IEEE, Kuala Lumpur, Malaysia, pp 238–243
23. Almomani I, Alkhayer A, El-Shafai W (2022) A crypto-steganography approach for hiding ransomware within hevc streams in android iot devices. Sensors 22(6):2281
24. Almomani I, AlKhayer A, El-Shafai W (2021) Novel ransomware hiding model using HEVC steganography approach. Comput Mater Contin 70(2):1209–1228
25. Choi SY, Lim CG, Kim YM (2019) Automated link tracing for classification of malicious websites in malware distribution networks. J Inf Process Syst 15(1):100–115
26. Kim D (2019) Potential risk analysis method for malware distribution networks. IEEE Access 7:185157–185167
27. Rudd EM, Rozsa A, Günther M, Boult TE (2016) A survey of stealth malware attacks, mitigation measures, and steps toward autonomous open world solutions. IEEE Commun Surv Tutorials 19(2):1145–1172
28. Kapoor A, Gupta A, Gupta R, Tanwar S, Sharma G, Davidson IE (2021) Ransomware detection, avoidance, and mitigation scheme: a review and future directions. Sustainability 14(1):8
29. Djenna A, Bouridane A, Rubab S, Marou IM (2023) Artificial intelligence-based malware detection, analysis, and mitigation. Symmetry 15(3):677

30. Gazzan M, Sheldon FT (2023) Opportunities for early detection and prediction of ransomware attacks against industrial control systems. Future Internet 15(4):144
31. Gorment NZ, Selamat A, Cheng LK, Krejcar O (2023) Machine learning algorithm for malware detection: taxonomy, current challenges and future directions. IEEE Access, 11:1–50
32. Samtani S, Zhao Z, Krishnan R (2023) Secure knowledge management and cybersecurity in the era of artificial intelligence. Inf Syst Front 25(2):425–429
33. Akhtar MS, Feng T (2023) Evaluation of machine learning algorithms for malware detection. Sensors 23(2):946

CONTENTS

CHAPTER 1

A Deep-Vision-Based Multi-class Classification System of Android Malware Apps

Iman Almomani, Walid El-Shafai, Mohanned Ahmed, Sara AlAnsary, Ghada AlMudahi, and Lama AlSwayeh

1.1 Introduction

Nowadays, smartphones have become an essential part of our lives because they are not used only for phone calls; they can be used for personal payment, keeping personal data, healthcare facilities, and other different personal services and applications [17]. Furthermore, it is commonly known that Android Operating System (OS) is considered the most

I. Almomani (✉)
Security Engineering Lab, Prince Sultan University, Riyadh, Saudi Arabia

Computer Science Department, The University of Jordan, Amman, Jordan
e-mail: imomani@psu.edu.sa; i.momani@ju.edu.jo

W. El-Shafai
Security Engineering Lab, Computer Science Department, Prince Sultan University, Riyadh, Saudi Arabia

I. Almomani et al. (eds.), *Cyber Malware*, Security Informatics and Law Enforcement, https://doi.org/10.1007/978-3-031-34969-0_1

popular OS for personal smartphones compared to other operating systems. Therefore, smartphones with Android OS platforms are highly targeted by cybercriminal operations [14].

Android OS is a Linux-based open-source OS developed and sponsored by Google company [23]. Google offers an authorized mobile play store with millions of Android applications (APKs). However, the Google play store is not the only open source of APKs; several other unauthorized third-party mobile stores are available for APKs. Therefore, due to the availability of a massive number of official and unofficial sources for APKs, the number of possible privacy and security problems by malicious software is boosted.

In 2017, Google company introduced a machine learning (ML)-based system called "Play Protect" as an attempt to alleviate the malware risks [24]. This ML-based ecosystem checks the APKs before and after their uploading to the Google market. However, it has been reported that over one million users have been infected by APKs available on the Google play store, and they have different types of malware [32]. Therefore, researchers and developers must put great effort into developing efficient and accurate malware detection tools to reduce the effect of growing malware risks.

Three different types of malware detection (MD) scenarios could be used to detect malware risks: static-based MD, dynamic-based MD, and vision-based MD [5, 7, 10]. The static-based MD algorithms analyze the Android APKs to extract some static features without running the APK files [6]. On the other hand, the dynamic-based MD algorithms monitor and examine the malware running behavior inside an isolated operating environment (i.e., an Android emulator) to check the behavior and effect of the produced malware traffic [8, 15]. Finally, the vision-based MD algorithms convert the Android APK files or their extracted features into

---

Electronics and Electrical Communication Engineering Department, Faculty of Electronic Engineering, Menoufia University, Menouf, Egypt
e-mail: welshafai@psu.edu.sa; walid.elshafai@el-eng.menofia.edu.eg

M. Ahmed • S. AlAnsary • G. AlMudahi • L. AlSwayeh
Security Engineering Lab, Computer Science Department, Prince Sultan University, Riyadh, Saudi Arabia
e-mail: mqasem@psu.edu.sa; 221421242@psu.edu.sa; 221420463@psu.edu.sa; 221421227@psu.edu.sa

visual images before in-depth training and testing analyses by different artificial intelligence (AI) tools, including machine learning (ML) and deep learning (DL) algorithms [12, 25].

The main advantages of vision-based MD algorithms compared to other static-based or dynamic-based MD algorithms are reducing the computational cost and avoiding reverse engineering steps required in static-based malware analysis [9]. In addition, they do not need to utilize special and isolated running environments to check the behavior of the malware APKs as required in dynamic-based malware analysis [11]. Thus, this encourages us to introduce a deep-vision-based multi-class classification system for android malware apps. The proposed vision-based MD system is based on utilizing 21 different convolutional neural network (CNN) models for malware detection and recognition. In the proposed MD system, the android apps' binary formats are first converted into color and grayscale vision formats before forwarding them to utilized CNN models for training and testing mechanisms. The classification performance of the proposed vision-based detection system is examined using different security and recognition metrics. The obtained results prove the high detection performance of the suggested MD system in effectively detecting various malware families.

The main contributions in this chapter are summarized as follows:

- Outlining the most recent related Android malware detection systems tested on the two open-source CICAndMal2017 and CICMalDroid2020 datasets.
- Proposing an accurate deep-vision-based multi-class classification system for Android malware apps for two recent Android datasets that compose different malware families.
- Developing and implementing different 21 fined-tuned DL algorithms on the proposed classification system.
- Testing the proposed classification system's performance on different color and grayscale vision formats of the android malware families.
- Presenting comprehensive security and detection analyses for the proposed classification system using different assessment parameters.

The remainder of this chapter is structured as follows. Section 1.2 presents the summary of the recent Android malware detection algorithms applied to the CICAnd- Mal2017 and CICMalDroid2020 datasets. Section 1.3 explains the proposed vision-based multi-class classification system. Section 1.4 introduces the evaluations and discussions. Finally, Sect. 1.5 provides the concluding remarks and future works.

1.2 Related Works

Different malware detection and classification models have been introduced in the literature for malware analysis of Android apps using two open-source CICAndMal2017 and CICMalDroid2020 datasets [26, 27]. To the best of our knowledge, all these malware analysis studies were based on static, dynamic, or hybrid-based detection techniques. But, none of these previous related studies are based on vision-based detection algorithms. Thus, this motivated us to introduce a deep-vision-based multi-class classification system for Android malware apps in the CICAndMal2017 and CICMalDroid2020 datasets.

Therefore, different recent static- or dynamic-based malware analysis and classification systems were introduced based on the CICAndMal2017 and CICMalDroid2020 datasets [2–4, 16, 16, 19, 21, 28–31, 34]. Mahshid et al. [19] used the CICAndMal2017 dataset with 89 extracted network traffic features to detect and categorize Android malware. They employed two deep neural models that were based on a hybrid of long short-term memory (LSTM) and convolution neural network (CNN). Their proposed CNN-LSTM model achieved a malware category classification with an accuracy of 98.9% and malware family classification with an accuracy of 97.29%.

Syed et al. [21] suggested a DeepAMD model based on different ML algorithms to identify and detect various android malware families. They tested their proposed model using the CICAndMal2017 and CICMalDroid2020 datasets, including benign and malicious Android apps. The introduced DeepAMD model was based on both dynamic and static detection algorithms. For the static-based analysis, the DeepAMD model achieved a 93.4% accuracy for binary malware classification, while it introduced 93.1% accuracy in the case of multi-malware families classification. In addition, for the dynamic analysis, the DeepAMD model scored a maximum detection accuracy of 80.3% for binary malware classification and 59.0% for a different malware families classification.

In [16], the authors introduced a detection algorithm based on the hybridization of gated recurrent units (GRU) and recurrent neural network

(RNN) for identifying malicious attacks in Android apps. First, they collected two control attributes from Android apps: program interface (API) phones and privileges. Then, they tested the proposed detection algorithm on the different Android families of the CICAndMal2017 dataset. The results reveal that the developed DL algorithm is better than several other detection methods by achieving a 98.2% detection accuracy.

In [31], the authors suggested an ML-based ransomware detection model. This model uses different ML algorithms to extract and analyze the valuable features of Android ransomware apps. They tested the detection efficacy of their proposed model on the CICMalDroid2020 dataset to check its classification accuracy of 10 distinct families of ransomware apps. The obtained results prove that the random forest classifier achieved superior ransomware detection efficiency compared to those of the other employed ML-based classifiers.

The authors in [4] presented an android malware detection system using five different ML algorithms and one DL algorithm to analyze and extract the main static features of the Android apps. These extracted features were permissions, API calls, permissions rate, and monitoring system events. The suggested detection system was examined using the CICAndMal2017 dataset that composes both benign and malware apps, and it achieved a detection accuracy of 98%.

In [30], a semi-supervised ML algorithm is presented to distinguish between ransomware from benign Android apps. The proposed algorithm composes different feature extraction and selection techniques that were tested on different labeled and unlabeled Android apps of the CICAndMal2017 dataset. In [34], a host-level encrypted traffic shaping-based Android malware classification approach was proposed. The classification approach tested three different ML algorithms on the real-world CICMalDroid2020 dataset for feature extraction and detection mechanisms. In addition, the authors simulated two experimental scenarios: malware family classification and binary malware detection. The results proved that the proposed classification approach had an accuracy of 98.8% for binary malware classification, while it achieved a 95.2% detection accuracy for malware family classification scenario.

In [2], a conversation-level traffic feature-based Android malware categorization and detection approach was presented. This approach consisted of four phases: feature extraction, data cleaning, feature selection, and training and testing. Different ML-based classifiers were tested, and the attained results tested on the CICMalDroid2020 dataset proved that the extra-trees ML classifier achieved superior detection and categorization efficiency in terms of accuracy, recall, and precision compared to those of other employed ML-based classifiers.

In [16], a novel DL-based Android malware detection algorithm was presented. It was based on utilizing a gated recurrent network to extract the static API and permission features from Android apps. The proposed detection algorithm was trained and tested on the CICAndMal2017 dataset, and an accuracy of 98.2% was accomplished. In [3], AI-based Android malware analysis and detection systems were introduced. This malware detection system was based on using different ML and DL algorithms such as k-nearest neighbors (KNN), support vector machine (SVM), autoencoder, long short-term memory (LSTM), linear discriminant analysis (LDA), and convolution neural network long short-term memory (CNN-LSTM) algorithms. The proposed detection system was tested using the CICMalDroid2020 dataset. The SVM achieved the highest detection accuracy of 99.5% for the detection analysis of the employed ML algorithms. For the security analysis of the DL algorithms, the LSTM algorithm presented a superior accuracy of 98.7%.

In [29], the authors presented different supervised learning algorithms for network traffic-based malware detection. Therefore, the Android malware detection was based on employing six algorithms: SVM, naïve Bayes, decision tree, multilayer perceptron neural network, K-nearest neighbors, and random forest. These ML-based classifiers were examined using the CICAndMal2017 dataset for binary and multi-classification scenarios. The random forest classifier introduced the best malware detection performance compared to other tested classifiers in terms of the obtained recall, precision, and accuracy results.

In [28], an efficient pseudo-label stacked auto-encoder (PLSAE)-based Android malware detection approach was suggested. The PLSAE algorithm is a semi-supervised learning approach; thus, it does not require more labeled data for efficient malware detection. The experimental analysis examines both labeled and unlabeled training Android instances. In addition, the feature extraction operation utilized both static and dynamic mechanisms. The security and detection analysis was based on analyzing five different Android categories of the CICMalDroid2020 dataset. The obtained outcomes proved that the suggested semi-supervising algorithm introduced high detection accuracy compared to other traditional supervised and semi-supervised ML algorithms.

1.3 Proposed Deep-Vision-Based Multi-class Classification System

The main aim of this chapter is to propose a deep vision-based multi-class classification system for Android malware apps. As shown in Fig. 1.1, the proposed vision-based classification system composes four consecutive phases: dataset gathering, preprocessing, training, and testing and evaluation.

In the first phase of dataset gathering, we collected the Android APKs and categorized them into different families, as presented in Table 1.1. In addition, in this phase, we ensured that there were no intersections or repeating apps in each tagged family. Then, in the second preprocessing phase, the Android apps were converted into two different vision-based formats (color and grayscale). The conversion process of Android apps to vision formats is performed by converting the 1D byte vectors of the portable executable Android files into 2D vector images [8, 15].

In the third phase of the proposed classification system, we performed the training process for the employed DL algorithms. First, we split the vision-based dataset into three different subsets: training, validation, and testing with the ratios of 80%, 10%, and 10%. Then, the training process starts by training each one of the utilized CNN algorithms by the 80% of generated vision-based Android images. Finally, the best-trained parameters of each employed CNN algorithm were collected and saved to be exploited in the testing stage. In the training process, we used two formats of color and gray images to extensively train the used CNN algorithms to examine their detection capabilities in identifying the Android malware attacks when they are represented in different vision-based formats.

In the proposed multi-class classification system, 21 different pre-trained CNN algorithms (ResNet50, VGG16, DenseNet121, VGG19, DenseNet201, DenseNet169, EfficientNetB7, EfficientNetB6, EfficientNetB5, EfficientNetB4, EfficientNetB3, EfficientNetB2, EfficientNetB1, EfficientNetB0, InceptionResNetV2, MobileNet, InceptionV3, MobileNetV3Large, MobileNetV2, Xception, and MobileNetV3Small) [13, 20, 33] were tested. These CNN algorithms were implemented in Python and developed by TensorFlow and Keras libraries [1, 18, 22].

Finally, in the last testing and evaluation phase, we exploited the best collected fine-tuned parameters resulting from the prior stage to accurately test and evaluate the utilized CNN algorithms. So, in this phase, we used

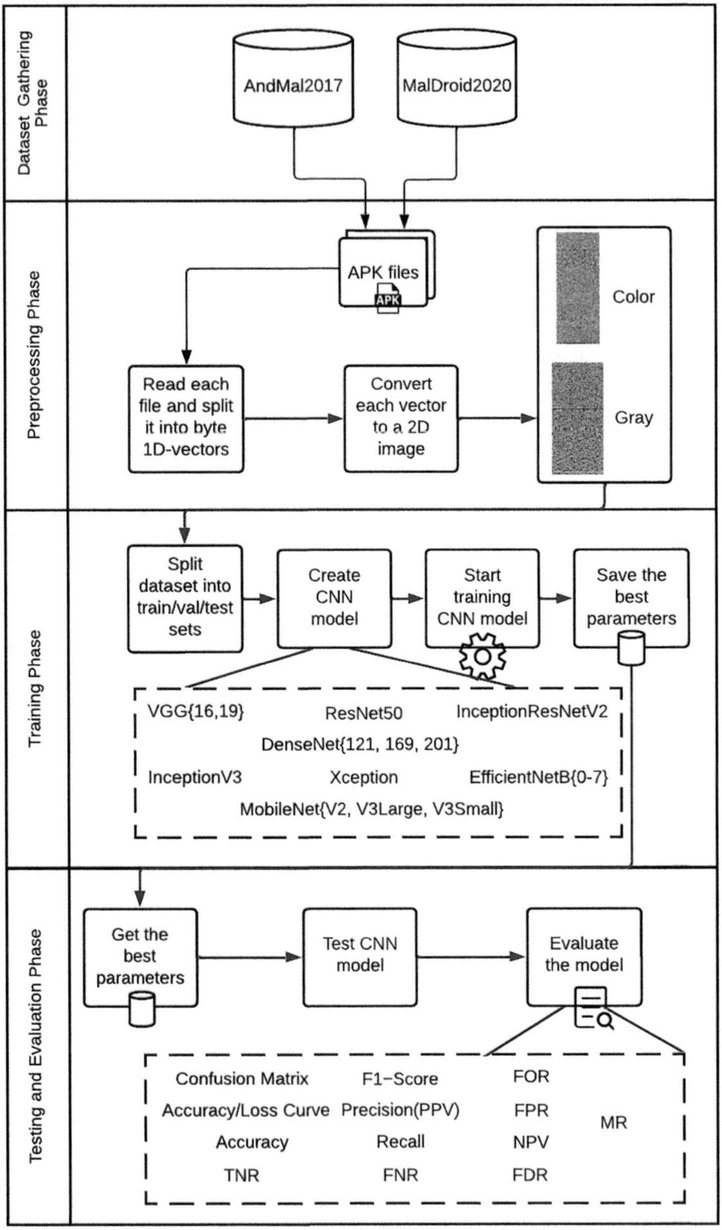

Fig. 1.1 Flow structure of the proposed deep-vision-based multi-class classification system

Table 1.1 Families and the number of samples of the employed Android datasets

Dataset	*Family*	*No. of samples*
CICAndMal2017	Benign	1648
	Adware	103
	Ransomware	100
	Scareware	112
	SMSMalware	109
CICMalDroid2020	Benign	4042
	Adware	1515
	Banking	2506
	Riskware	4362
	SMSMalware	4822

10% of the visual Android images for the validation process and 10% for the testing process. In the evaluation of the employed CNN algorithms, different security and detection assessment parameters are estimated, as will be discussed and clarified in Sect. 1.4.3.

1.4 Evaluations and Discussions

This section presents a discussion of the security analysis and detection assessment of the proposed vision-based multi-class classification system for Android malware apps. All experimental tests were carried out using the vision formats of the Android apps included in the CICAndMal2017 [26] and CICMalDroid2020 [27] datasets, as clarified in Sect. 1.4.1. The proposed vision-based malware detection system is extensively evaluated using different detection and evaluation metrics, as indicated in Sect. 1.4.2.

1.4.1 Datasets Description

We tested the detection performance of the proposed multi-class classification system on the vision-based formats of two standard unbalanced Android datasets: CICAndMal2017 and CICMalDroid2020 [26, 27]. Table 1.1 presents the names of the families and the number of samples included in the examined Android datasets. As indicated in Table 1.1, each one of these datasets consists of five families: one benign and four malware families. For example, the first CICAndMal2017 dataset composes of Adware, Ransomware, Scareware, and SMSmalware malicious families, while the second CICMalDroid2020 dataset composes of Adware, Banking, Riskware, and SMSMalware families.

Table 1.2 shows samples of the vision-based color and grayscale formats of the binary Android APKs for the two examined datasets. It is observed that each android family, benign or malware, has distinct features and characteristics compared to other android families. So, vision-based Android malware detection models are highly recommended for malware classification systems compared to other static or dynamic detection models.

1.4.2 Security Detection Metrics

In the performance analysis of the proposed vision-based multi-class classification system, various security detection metrics were employed to comprehensively test the detection and classification efficacy of the examined DL algorithms. The mathematical formulas of the exploited security detection metrics are expressed as follows:

$$\text{F1-Score} = \frac{2TP}{2TP + FN + FP} \tag{1.1}$$

$$\text{Precision (PPV)} = \frac{TP}{FP + TP} \tag{1.2}$$

$$\text{Recall} = \frac{TP}{TP + FN} \tag{1.3}$$

$$\text{Accuracy} = \frac{TN + TP}{TN + FP + TP + FN} \tag{1.4}$$

$$\text{Misclassification rate (MR)} = \frac{FP + FN}{TN + FP + TP + FN} \tag{1.5}$$

$$\text{FOR} = \frac{FN}{FN + TN} \tag{1.6}$$

$$\text{FDR} = \frac{FP}{FP + TP} \tag{1.7}$$

$$\text{FNR} = \frac{FN}{FN + TP} \tag{1.8}$$

Table 1.2 Vision-based color and grayscale samples of Android APKs

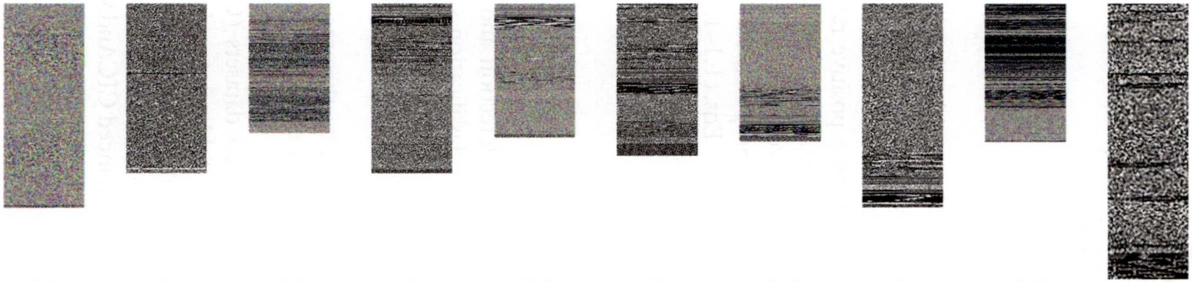

Color	Gray	Color	Gray	Color	Gray	Color	Gray	Color	Gray
Benign		Adware		Ransomware		Scareware		SMSMalware	

(i) CICAndMal2017 dataset.

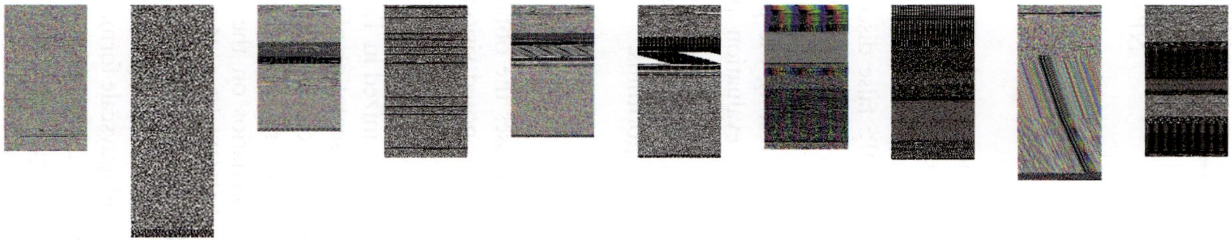

Color	Gray	Color	Gray	Color	Gray	Color	Gray	Color	Gray
Benign		Adware		Banking		Riskware		SMSMalware	

(ii) CICMalDroid2020 dataset.

$$\mathrm{FPR} = \frac{FP}{FP + TN} \tag{1.9}$$

$$\mathrm{NPV} = \frac{TN}{FN + TN} \tag{1.10}$$

$$\mathrm{TNR} = \frac{TN}{FP + TN}, \tag{1.11}$$

where the confusion matrix of each examined DL algorithm is exploited to estimate the numerical values of the *TN* (true negative), *TP* (true positive), *FN* (false negative), and *FP* (false positive) parameters. Where, FOR: false omission rate, FDR: false discovery rate, TPR: true positive rate, FNR: false negative rate, FPR: false positive rate, NPV: negative predictive value, TNR: true negative rate, and PPV: positive predictive value.

In addition to testing the performance of the proposed classification system using the evaluation parameters given in Eqs. (1.1–1.11), the confusion matrix [15] and loss and accuracy curves [8] were also utilized to comprehensively assess the security and detection capabilities of the employed CNN algorithms.

1.4.3 Results Analysis

This section discusses the obtained results of the detection and security analysis for the proposed vision-based multi-class classification system in terms of the assessment parameters described in Sect. 1.4.2. The experimental parameters of the CNN algorithms used in the proposed classification system are organized in Table 1.3.

For a simple presentation of the analysis of the results, the precision, F1-Score, recall, and accuracy metrics are emphasized for each employed CNN algorithm in the proposed classification system for color and grayscale vision-based scenarios on the examined unbalanced datasets (CICAndMal2017 and CICMalDroid2020), as revealed in Tables 1.4 and 1.5.

Tables 1.4 and 1.5 introduce the detection and security performance analysis of all 21 CNN predictive algorithms examined on two vision-based color and grayscale formats of the two unbalanced CICAndMal2017 and CICMalDroid2020 datasets. The attained results disclosed that the MobileNetV3Large and VGG16 CNN algorithms achieve superior detection efficiency for the color and grayscale formats of the CICAndMal2017

Table 1.3 Experimental parameters of the CNN algorithms used in the proposed classification system

Simulation parameter	*Value*
Software Python libraries	TensorFlow and Keras
Training/testing/validation ratio	80/10/10 (%)
CNN optimizer	ADAM
Learning rate	0.0001
Regularizer decay rate	0.001
CNN regularizer	L2 regularizer algorithm
Epochs number	128
Function of loss	Categorical cross-entropy function
Minimum batch size	64

dataset, respectively. In addition, the obtained results clarified that the EfficientNetB7 and ResNet50 CNN algorithms accomplish the most excellent detection proficiency for the color and grayscale formats of the CICMalDroid2020 dataset, respectively.

To provide a simple presentation of the experimental analysis, we show only the confusion matrices and loss and accuracy curves of the best-accomplished CNN algorithms for the two examined vision-based color and grayscale formats of the two unbalanced datasets. Figure 1.2 shows the obtained confusion matrices of the best-performed MobileNetV3Large and VGG16 CNN algorithms for the two vision-based color and grayscale formats of the CICAndMal2017 dataset. Figure 1.3 demonstrates the attained confusion matrices of the best-performed EfficientNetB7 and ResNet50 CNN algorithms for the two vision-based color and grayscale formats of the CICMalDroid2020 dataset. It is revealed from these obtained confusion matrices that the MobileNetV3Large, VGG16, EfficientNetB7, and ResNet50 CNN algorithms provide low false detection performance for the two vision-based color and grayscale formats in both CICAndMal2017 & CICMalDroid2020 datasets.

Figure 1.4 presents the obtained accuracy and loss curves of the best-accomplished MobileNetV3Large and VGG16 CNN algorithms for the two vision-based color and grayscale formats of the CICAndMal2017 dataset. Figure 1.5 presents the obtained accuracy and loss curves of the best-executed EfficientNetB7 and ResNet50 CNN algorithms for the two vision-based color and grayscale formats of the CICMalDroid2020 dataset. The acquired simulation outcomes verify that the MobileNetV3Large, VGG16, EfficientNetB7, and ResNet50 CNN algorithms offer the lowest

Table 1.4 Detection performance analysis of the proposed classification system for the CICAndMal2017 dataset

Model	*Format*	*Accuracy (%)*	*F1-score (%)*	*Precision (%)*	*Recall (%)*
VGG16	Color	86.12	84.41	84.87	86.12
	Gray	89.0	88.16	88.5	89.0
ResNet50	Color	86.6	85.72	86.97	86.6
	Gray	87.56	86.73	88.97	87.56
VGG19	Color	86.6	85.89	87.52	86.6
	Gray	88.04	87.6	87.83	88.04
DenseNet121	Color	86.12	83.77	81.85	86.12
	Gray	84.69	82.78	86.66	84.69
DenseNet169	Color	85.17	83.37	85.84	85.17
	Gray	87.08	86.02	85.82	87.08
DenseNet201	Color	87.08	85.82	87.29	87.08
	Gray	86.6	85.08	86.04	86.6
EfficientNetB0	Color	84.21	81.33	83.26	84.21
	Gray	84.21	81.09	81.93	84.21
EfficientNetB1	Color	86.6	85.54	85.32	86.6
	Gray	84.21	81.82	80.57	84.21
EfficientNetB2	Color	85.17	84.09	84.62	85.17
	Gray	85.65	84.17	84.24	85.65
EfficientNetB3	Color	87.08	85.69	88.47	87.08
	Gray	86.6	85.7	86.51	86.6
EfficientNetB4	Color	87.08	85.86	85.52	87.08
	Gray	86.6	86.23	86.11	86.6
EfficientNetB5	Color	84.21	82.12	84.27	84.21
	Gray	86.12	84.58	86.01	86.12
EfficientNetB6	Color	82.3	78.59	80.42	82.3
	Gray	83.25	80.07	86.12	83.25
EfficientNetB7	Color	84.21	81.52	81.89	84.21
	Gray	81.82	78.49	77.35	81.82
InceptionResNetV2	Color	79.9	74.9	71.1	79.9
	Gray	78.95	69.66	62.33	78.95
InceptionV3	Color	83.25	79.88	78.89	83.25
	Gray	83.73	81.27	81.02	83.73
MobileNet	Color	81.82	78.17	76.63	81.82
	Gray	82.3	79.69	80.59	82.3
MobileNetV2	Color	81.34	77.64	75.66	81.34
	Gray	82.3	79.2	78.41	82.3
MobileNetV3Large	Color	86.6	86.46	87.69	86.6
	Gray	88.04	87.21	88.22	88.04
MobileNetV3Small	Color	85.17	84.68	84.6	85.17
	Gray	85.65	83.68	85.13	85.65
Xception	Color	80.38	78.88	78.0	80.38
	Gray	82.3	80.52	80.72	82.3

Table 1.5 Detection performance analysis of the proposed classification system for the CICMalDroid2020 dataset

Model	Format	*Accuracy (%)*	*F1-score (%)*	*Precision (%)*	*Recall (%)*
VGG16	Color	88.6	88.64	88.72	88.6
	Gray	89.29	89.29	89.31	89.29
ResNet50	Color	88.77	88.82	88.89	88.77
	Gray	89.99	89.99	90.03	89.99
VGG19	Color	88.66	88.77	88.94	88.66
	Gray	88.66	88.69	88.77	88.66
DenseNet121	Color	87.04	86.95	87.1	87.04
	Gray	85.71	85.83	86.23	85.71
DenseNet169	Color	86.17	86.26	86.89	86.17
	Gray	86.92	86.75	86.67	86.92
DenseNet201	Color	87.15	87.2	87.4	87.15
	Gray	88.66	88.63	88.66	88.66
EfficientNetB0	Color	89.64	89.67	89.72	89.64
	Gray	89.24	89.18	89.15	89.24
EfficientNetB1	Color	89.7	89.72	89.77	89.7
	Gray	89.47	89.58	89.8	89.47
EfficientNetB2	Color	88.02	88.03	88.05	88.02
	Gray	88.6	88.54	88.63	88.6
EfficientNetB3	Color	88.89	88.78	89.03	88.89
	Gray	89.0	88.97	89.1	89.0
EfficientNetB4	Color	89.64	89.68	89.74	89.64
	Gray	88.77	88.83	88.9	88.77
EfficientNetB5	Color	87.96	88.05	88.4	87.96
	Gray	89.41	89.41	89.41	89.41
EfficientNetB6	Color	88.19	88.16	88.16	88.19
	Gray	87.62	87.63	87.66	87.62
EfficientNetB7	Color	89.87	89.88	90.01	89.87
	Gray	88.37	88.51	88.88	88.37
InceptionResNetV2	Color	54.8	47.81	46.43	54.8
	Gray	60.42	54.69	57.34	60.42
InceptionV3	Color	81.19	80.86	81.03	81.19
	Gray	80.61	79.81	80.4	80.61
MobileNet	Color	78.18	77.84	77.72	78.18
	Gray	83.8	83.91	84.13	83.8
MobileNetV2	Color	79.92	80.0	80.2	79.92
	Gray	83.97	83.82	83.87	83.97
MobileNetV3Large	Color	89.41	89.4	89.47	89.41
	Gray	89.7	89.67	89.72	89.7
MobileNetV3Small	Color	88.31	88.27	88.23	88.31
	Gray	88.66	88.68	88.82	88.66
Xception	Color	80.56	81.13	82.57	80.56
	Gray	82.47	82.49	82.66	82.47

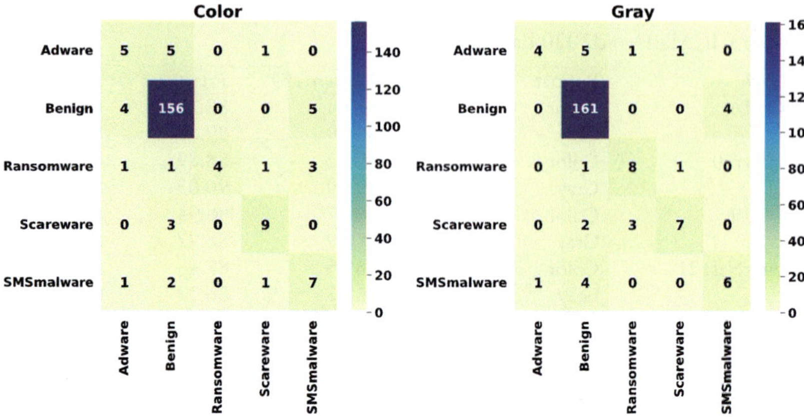

Fig. 1.2 Confusion matrices of the best-performed CNN algorithms on the color and grayscale formats of the CICAndMal2017 dataset. (**a**) MobileNetV3Large. (**b**) VGG16

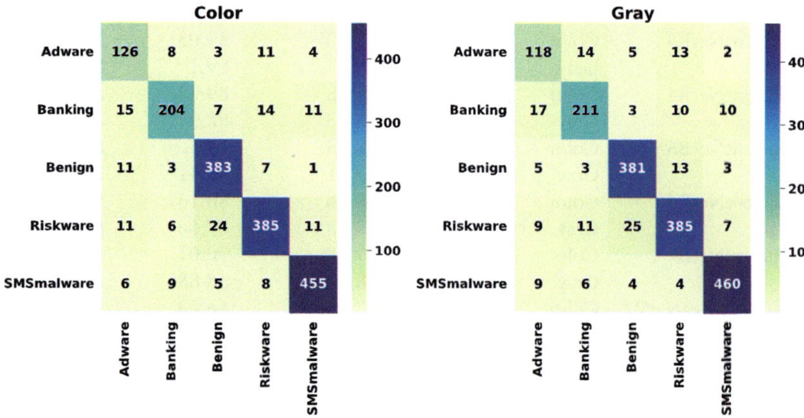

Fig. 1.3 Confusion matrices of the best-performed CNN algorithms on the color and grayscale formats of the CICMalDroid2020 dataset. (**a**) EfficientNetB7. (**b**) ResNet50

detection loss and highest detection accuracy compared to the other tested CNN algorithms used in the proposed classification system, as also summarized in Tables 1.4 and 1.5.

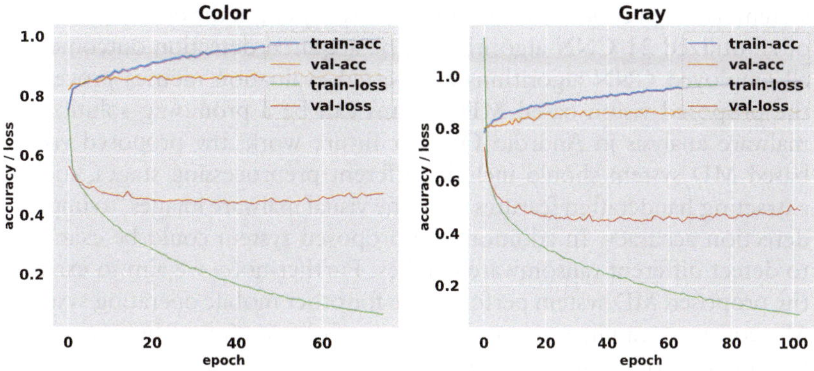

Fig. 1.4 Accuracy and loss curves for the best-performed CNN algorithms on the color and grayscale formats of the CICAndMal2017 dataset. (**a**) MobileNetV3Large. (**b**) VGG16

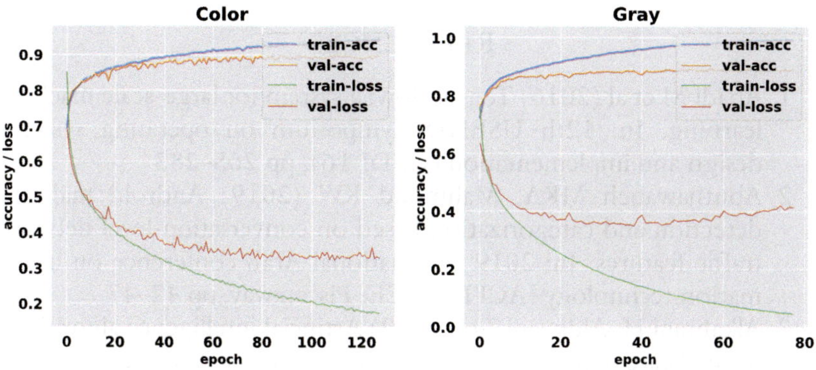

Fig. 1.5 Accuracy and loss curves for the best-performed CNN algorithms on the color and grayscale formats of the CICMalDroid2020 dataset. (**a**) EfficientNetB7. (**b**) ResNet50

1.5 Conclusions and Future Work

This chapter introduced an efficient vision-based multi-classification system to detect different types of malware families in Android apps. The proposed MD system could be adapted to detect malware in Android apps in visual color or grayscale formats. An in-depth evaluation of the proposed MD system has been conducted to comprehensively check the detection efficacy of all utilized 21 CNN algorithms. The acquired detection outcomes for all employed CNN algorithms and tested evaluation metrics prove that the proposed vision-based MD systems can be a promising solution for malware analysis in Android OS. For future work, the proposed vision-based MD system should include different preprocessing stages, such as extracting handcrafted features from the visual malware images, to improve detection accuracy. In addition, the proposed system could be examined to detect different ransomware families. Furthermore, we aim to examine the proposed MD system performance for other mobile operating systems. Moreover, we can investigate the impact of obfuscation techniques on the performance of vision-based MD systems.

Acknowledgments The authors would like to thank the support of Prince Sultan University. Moreover, this research was done during the author Iman Almomani's sabbatical year 2021/2022 from the University of Jordan, Amman, Jordan.

References

1. Abadi M et al (2016) TensorFlow: a system for large-scale machine learning. In: 12th USENIX symposium on operating systems design and implementation (OSDI 16), pp 265–283
2. Abuthawabeh MKA, Mahmoud KW (2019) Android malware detection and categorization based on conversation-level network traffic features. In: 2019 International Arab conference on information technology (ACIT). IEEE, Piscataway, pp 42–47
3. Alkahtani H, Aldhyani TH (2022) Artificial intelligence algorithms for malware detection in android-operated mobile devices. Sensors 22(6):2268
4. Almahmoud M, Alzu'bi D, Yaseen Q (2021) Redroiddet: android malware detection based on recurrent neural network. Proc Comput Sci 184:841–846
5. Almohaini R, Almomani I, AlKhayer A (2021) Hybrid-based analysis impact on ransomware detection for android systems. Appl Sci 11(22):10976
6. Almomani I, AlKhayer A, Ahmed M (2021) An efficient machine learning-based approach for android v. 11 ransomware detection. In: 2021 1st international conference on artificial intelligence and data analytics (CAIDA). IEEE, Piscataway, pp 240–244

7. Almomani I, Qaddoura R, Habib M, Alsoghyer S, Al Khayer A, Aljarah I, Faris H (2021) Android ransomware detection based on a hybrid evolutionary approach in the context of highly imbalanced data. IEEE Access 9:57674–57691
8. Almomani I, Alkhayer A, El-Shafai W (2022) An automated vision-based deep learning model for efficient detection of android malware attacks. IEEE Access 10:2700–2720
9. Awan MJ, Masood OA, Mohammed MA, Yasin A, Zain AM, Damaševičius R, Abdulkareem KH (2021) Image-based malware classification using VGG19 network and spatial convolutional attention. Electronics 10(19):2444
10. Bakour K, Ünver HM (2021) Visdroid: android malware classification based on local and global image features, bag of visual words and machine learning techniques. Neural Comput Appl 33(8):3133–3153
11. Ben Abdel Ouahab I, Elaachak L, Bouhorma M (2022) Classification of malicious and benign binaries using visualization technique and machine learning algorithms. In: Big data intelligence for smart applications. Springer, Berlin, pp 297–315
12. Bovenzi G, Cerasuolo F, Montieri A, Nascita A, Persico V, Pescapé A (2022) A comparison of machine and deep learning models for detection and classification of android malware traffic. In: IEEE DistInSys 22
13. Brownlee J (2016) Deep learning with Python: develop deep learning models on Theano and TensorFlow using Keras
14. Dhalaria M, Gandotra E (2021) Android malware detection techniques: a literature review. Recent Patents Eng 15(2):225–245
15. El-Shafai W, Almomani I, AlKhayer A (2021) Visualized malware multi-classification framework using fine-tuned CNN-based transfer learning models. Appl Sci 11(14):6446
16. Elayan ON, Mustafa AM (2021) Android malware detection using deep learning. Proc Comput Sci 184:847–852
17. Garg S, Baliyan N (2021) Comparative analysis of android and IOS from security viewpoint. Comput Sci Rev 40:100372
18. Géron A (2019) Hands-on machine learning with Scikit-Learn, Keras, and TensorFlow: concepts, tools, and techniques to build intelligent systems

19. Gohari M, Hashemi S, Abdi L (2021) Android malware detection and classification based on network traffic using deep learning. In: 2021 7th international conference on web research (ICWR). IEEE, Piscataway, pp 71–77
20. Hodnett M, Wiley JF (2018) R Deep Learning Essentials: a step-by-step guide to building deep learning models using TensorFlow, Keras, and MXNet
21. Imtiaz SI, ur Rehman S, Javed AR, Jalil Z, Liu X, Alnumay WS (2021) Deepamd: Detection and identification of android malware using high-efficient deep artificial neural network. Fut Gener Comput Syst 115:844–856
22. Joseph FJJ, Nonsiri S, Monsakul A (2021) Keras and tensorflow: a hands-on experience, pp 85–111
23. Kiran Kumar M, Kranthi Kumar S, Kalpana E, Srikanth D, Saikumar K (2022) A novel implementation of Linux based android platform for client and server. In: A fusion of artificial intelligence and internet of things for emerging cyber systems. Springer, Berlin, pp 151–170
24. Kouliaridis V, Kambourakis G (2021) A comprehensive survey on machine learning techniques for android malware detection. Information 12(5):185
25. Kumar S, Janet B, Neelakantan S (2022) Identification of malware families using stacking of textural features and machine learning. Expert Syst Appl 208:118073
26. Lashkari AH, Kadir AFA, Taheri L, Ghorbani AA (2018) Toward developing a systematic approach to generate benchmark android malware datasets and classification. In: 2018 international Carnahan conference on security technology (ICCST). IEEE, Piscataway, pp 1–7
27. Mahdavifar S, Kadir AFA, Fatemi R, Alhadidi D, Ghorbani AA (2020) Dynamic android malware category classification using semi-supervised deep learning. In: 2020 IEEE international conference on dependable, autonomic and secure computing. International conference on pervasive intelligence and Computing, International conference on cloud and big data computing, International conference on cyber science and technology congress (DASC/PiCom/CBDCom/CyberSciTech). IEEE, Piscataway, pp 515–522

28. Mahdavifar S, Alhadidi D, Ghorbani A et al (2022) Effective and efficient hybrid android malware classification using pseudo-label stacked auto-encoder. J Netw Syst Manag 30(1):1–34
29. Manzano C, Meneses C, Leger P, Fukuda H (2022) An empirical evaluation of supervised learning methods for network malware identification based on feature selection. Complexity 18:6760920. https://doi.org/10.1155/2022/6760920
30. Noorbehbahani F, Saberi M (2020) Ransomware detection with semi-supervised learning. In: 2020 10th international conference on computer and knowledge engineering (ICCKE). IEEE, Piscataway, pp 024–029
31. Noorbehbahani F, Rasouli F, Saberi M (2019) Analysis of machine learning techniques for ransomware detection. In: 2019 16th international ISC (Iranian Society of Cryptology) conference on information security and cryptology (ISCISC). IEEE, Piscataway, pp 128–133
32. Razgallah A, Khoury R, Hallé S, Khanmohammadi K (2021) A survey of malware detection in android apps: recommendations and perspectives for future research. Comput Sci Rev 39:100358
33. Vasilev I, Slater D, Spacagna G, Roelants P, Zocca V (2019) Python Deep Learning: exploring deep learning techniques and neural network architectures with Pytorch, Keras, and TensorFlow
34. Zhou J, Niu W, Zhang X, Peng Y, Wu H, Hu T (2020) Android malware classification approach based on host-level encrypted traffic shaping. In: 2020 17th international computer conference on wavelet active media technology and information processing (ICCWAMTIP). IEEE, Piscataway, pp 246–249

CHAPTER 2

Android Malware Detection Based on Network Analysis and Federated Learning

Djallel Hamouda, Mohamed Amine Ferrag, Nadjette Benhamida, Zine Eddine Kouahla, and Hamid Seridi

2.1 Introduction

With the development and the increasing number of available Android-based systems and application software, such as in industrial IoT systems and smartphones [2], the latter are also becoming more popular targets for cyber criminals, who plant their malicious apps as an exploit to conduct serious and devastating cyber attacks over a large network of connected

D. Hamouda • N. Benhamida • Z. E. Kouahla • H. Seridi
Labstic Laboratory, Department of Computer Science, Guelma University, Guelma, Algeria
e-mail: hamouda.djallel@univ-guelma.dz; benhamida.nadjette@univ-guelma.dz; kouahla.zineeddine@univ-guelma.dz; seridi.hamid@univ-guelma.dz

M. A. Ferrag (✉)
Technology Innovation Institute, Abu Dhabi, United Arab Emirates
e-mail: mohamed.ferrag@tii.ae

I. Almomani et al. (eds.), *Cyber Malware*, Security Informatics and Law Enforcement, https://doi.org/10.1007/978-3-031-34969-0_2

devices. Several malware detection strategies have been proposed in the literature and consist primarily of two stages: malware analysis and malware detection (Fig. 2.1) [18]. The former describes the analysis and processing techniques used for detection and consists of two types of approaches: static analysis and dynamic analysis (Fig. 2.1a). The static analysis investigates the malware code without running it, using reverse engineering techniques. Usually, the application package (APK) file is decompressed, and many representative features are extracted from it [19]. This static analysis had proven to be effective against known malware. It is, however, ineligible with new ones and easily deceived by obfuscation techniques. Dynamic analysis monitor and analyze the runtime characteristics of malware applications while running their code, based on an assessment of behaviors to determine the functionality of malware, such as information flow tracking, function call monitoring, and instruction tracing, which can be applied [16]. Dynamic analysis generally uses virtual environments and emulators for analysis and data collection, an effective technique to identify unknown malware. However, it is time-consuming and requires costly computation. Dynamic malware detection was also introduced from a network traffic aspect to detect malware that conducts its attacks over networks to remote targets [4]. And due to this, they produce network traffic traces that can be detected by analyzing the network traffic behavior. However, the latter is also a tough task and suffers from increasing false alarm rates or decreasing sensitivity (i.e., detecting attack classes) as the number of different types of behaviors increases.

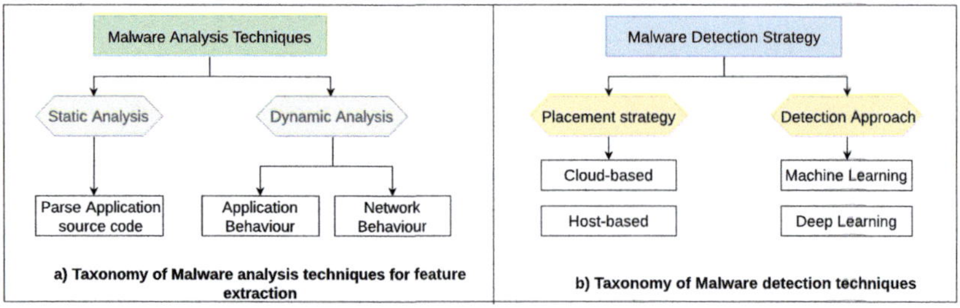

Fig. 2.1 A taxonomy of malware analysis techniques and detection strategies. (**a**) Taxonomy of Malware analysis techniques for feature extraction. (**b**) Taxonomy of Malware detection techniques

The malware detection strategy (Fig. 2.1b) describes the placement strategy and the detection approach. The placement strategy specifies whether the malware detection system is implemented on the host (a mobile or IoT device) or in the cloud to characterize the detection system's efficiency against complex code variants of malware while using limited computation resources [5]. On the other side, malware detection approaches describe the methods and algorithms used to detect and identify malware. Several ML and DL approaches have been proposed in this context to improve the accuracy of malware detection. However, their efficiency depends on the availability of large and diverse datasets. In addition to the described challenges, data privacy and shortages are another important challenge when deploying conventional cloud-based and deep-learning-based (DL) security solutions. In this chapter, we study the effectiveness of the recently proposed federated learning paradigm for malware detection against the aforementioned challenges. Our objective is to develop an efficient and effective learning mechanism for malware detection in terms of detection accuracy and computation cost, all while maintaining data privacy. The main contributions to this chapter are as follows:

- We first reviewed Android malware detection strategies and related challenges.
- We propose a novel privacy-preserving federated deep learning technique using convolutional neural networks (CNN) to detect several types of malware, based on network traffic behavior analysis.
- We evaluate the performance of the proposed detection methodology in terms of accuracy, detection rate, and under different FL settings (i.e., number of participating devices).

The remainder of this chapter is organized as follows. We review malware detection strategies in Sect. 2.1. Section 2.2 provides an overview of related works. Section 2.3 demonstrates the deployment architecture. Section 2.4 demonstrates experimental results and effectiveness of the proposed FL-based malware detection methodology. Finally, we conclude our work in Sect. 2.5.

2.2 Related Studies

Several studies have been conducted in the domain of Android malware detection relying on deep learning methodologies. Researches are making use of both static and dynamic malware analysis techniques to extract representative features necessary for the learning process of deep learning models. In a case study proposed by Yuan et al. [21] that links the features from the static analysis with the features from the dynamic analysis of Android apps for malware detection, the extracted features fall into three categories: required permission, sensitive API, and dynamic behavior. The authors proposed a deep learning approach based on a deep belief network (DBN) for malware detection engine (DroidDetector); the results reveal that the proposed model obtained improved accuracy under diverse scenarios and outperformed traditional machine learning techniques especially with the availability of more training data. Similar in [22], the authors used FlowDroid static analysis tool to extract data flows from all sensitive sources to sensitive sinks and proposed deep learning-based approach using DBN for identifying malware directly from their data flows. Results show that the proposed approach significantly outperforms traditional ML approaches in appropriate settings. Karbab et al. [10] presented an automatic detection system of Android malware using deep learning techniques and raw sequences of API method calls to identify Android malware. Kim et al. [11] proposed a multimodal deep learning method for malware detection based on only static analysis features by analyzing APK files such as manifest file, adex file, and a .so file to reflect various characteristics of applications in various aspects. The proposed deep learning multimodal was utilized to discriminate the properties of different types of input features that are processed in different initial networks separately, and the results of the initial networks are subsequently used to train the final network, to produce the classification results. Iadarola et al. [9] proposed a deep convolutional network (CNN) for image-based malware classification aimed to discriminate between Android malware and trusted samples (Benign). They also provided a methodology about the interpretability of the predictions performed by the model using a cumulative heatmap manually performed by the analyst. Experimental results demonstrated the efficiency of the proposed method, by identifying six different malware families from benign samples and also by providing interpretability about the predictions performed by the model. Unlike in [4], Arora et al. proposed the first Android malware detection using its network traffic analysis. They captured network traffic data of 13 malware apps. Then, dynamic analysis was conducted on the traffic behavior to generate significant features for malware detection. The authors selected a rule-based machine learning classifier and obtained reasonably acceptable results. Similar in [6], the authors presented a network-based detection model to distinguish malicious applications from normal ones installed on Android mobile devices. They tried various ML models to select the best deployment model. Another study [23] employed a dynamic detection approach based on network traffic, which captured the application's behavior throughout runtime. The authors considered seven network traffic features and obtained acceptable results. In [12], and [13], Lashkari et al. proposed new Android malware datasets using network traffic analysis. The authors installed a large number of Android apps (both benign and malicious) on real devices and then captured network traffic from user interactions to collect all normal and abnormal behaviors that characterize malware apps. The authors proposed several ML-based techniques to detect and identify malware apps. Experimental results showed the effectiveness of network-based malware detection. However, results for malware identification were not good enough. Recently, Rey et al. [17] investigated the potential afforded by the emergent federated learning (FL) paradigm for IoT malware detection, and the experimental findings indicated the promising performance with an extra security service, which is data privacy protection. Table 2.1 summarizes related works on Android malware detection based on deep learning.

In summary, several Android malware detection studies have been proposed and discussed, applying both ML and DL approaches and employing different malware analysis techniques with matching features. However, DL-based malware detection using the predictability of their network behavior has not been widely discussed. In addition, other constraints identified, including computing resource limitations, a lack of training data, and privacy concerns, have not been commonly discussed.

To this end, we proposed a novel privacy-preserving DL-based malware detection system employing the emergent federated learning paradigm to efficiently and effectively detect a large number of Android malware samples based on network analysis and without data sharing.

2.3 Methodology

2.3.1 Federated Learning Paradigm

Recently, a novel collaborative learning paradigm and a decentralized optimization strategy named federated learning (FL) have been proposed to train ML and DL models based on datasets and computational resources

Table 2.1 Summary of related studies of Android malware detection using DL

Reference	*Input features*	*D. approach*	*Performance %*	*Year*	*Dataset*
Yuan et al. [21]	Static and Dynamic	DBN	Acc = 96.76	2016	Contagio
Zhu et al. [22]	Static	DBM	Acc = 95.05	2017	MalGenome
Kim et al. [11]	Static	DNN	Acc = 85, Fnr = 14 Tnr = 96, Tpr = 85 Fpr = 15	2018	MalGenome
Karbab et al. [10]	Static and Dynamic	ANN	F1 = 98.18, Fpr = 1.15	2018	MalDozer
Garg et al. [6]	Dynamic	Random forest Decision tree K nearest neighbor,	Acc = 98 Fpr = 1.6	2017	N/A
Zulkifli et al. [23]	Dynamic	Decision tree	Acc = 97.6–98.4 Roc = 93.2–95.4	2018	Drebin Contagiodumpset
Iadarola et al. [9]	Static	CNN	Acc = 98.0, Pr = 97.2 Auc = 99.5, Rc = 97 F1 = 97.1	2021	Drebin MalGenome
Andresini et al. [3]	Dynamic	AutoEncoder	Acc = 89.63 Rc = 66.4–95.4 F1 = 71.9	2021	AAGM2017
Rey et al. [17]	Dynamic	MLP, AutoEncoder	Acc = 99–99.9 Tpr = 97–99 Tnr = 91–99	2022	N-BaIoT

Acc: Accuracy, **Pr**: Precision, **Rc**: Recal, **Tnr**: True negative rate, **Fpr** : False positive rate, **F1** : F2-score

that are distributed across multiple devices through parameter exchange while preventing data leakage [15]. This novel paradigm has the potential to overcome significant challenges of traditional ML approaches for malware detection, such as resource constraints, data privacy, data distribution, and heterogeneity [7]. The main idea of FL is to solve the conventional optimization problem of ML iteratively in a distributed manner, such that each device tries to minimize its local cost function through local training while seeking to optimize the global model parameters. The FL optimization problem can be formulated as follows:

$$\min_{W \in \mathbb{R}^d} f(W) = \frac{1}{N} \sum_{i=1}^{N} \beth(W, D_i) \tag{2.1}$$

where $W \in \mathbb{R}^d$ denotes the global model parameters to be optimized, N, D_i denotes the participating devices and their corresponding data samples, and $\beth(W, D_i)$ denotes the cost function to be optimized and returns the locally computed updates. In each round of FL, three main steps are performed [15]:

- Device sampling: A subset of devices, also called clients, is chosen according to selection criteria to participate in the training procedure.
- Local computation: Each chosen device trains its local model on its local dataset, minimizing its local cost function $\beth(W, D_i)$.
- Aggregation and consensus: The locally computed model updates are aggregated to update the global model, either with the help of a central entity called an aggregation server, which describes the centralized FL, or by communicating with only neighboring devices, which describes the decentralized FL.

Although FL provides data privacy and effective deployment of ML approaches, there are also research efforts to make FL more secure against inherited vulnerabilities within the framework, such as poisoning attacks, model stealing, and Byzantine attacks [8].

2.3.2 *Our Proposed Detection Methodology*

With the aim to efficiently and effectively detect large-scale Android malware while considering privacy preservation, we incorporate FL for model training. Our methodology consists mainly of three steps: data processing, FDL-based model training, and results evaluation (in Sect. 2.4).

Dataset Processing
Deep-learning-based malware detection is largely influenced by the quantity and quality of training data; the more high-quality data there is, the higher the accuracy and results. In this study, we selected an adequate and benchmark dataset named AAGM (Android Adware and General Malware) due to its diversity in malware samples [12]. This dataset comprises 1500 benign app samples and 400 malware samples from 10 families, including 5 families of adware and 5 families of general malware. The authors installed these samples on actual smartphones and started running user-interaction scenarios to capture meaningful network traffic behavior. They provided a total of 471597 benign instances and 160358 malware instances, along with 80 network traffic features (i.e., flow-based, time-based, and packet-based features), in order to distinguish Android malware behavior from that of benign apps. Before training, it is essential to do exploratory analyses and data processing on the dataset to handle multiple issues. First, we eliminate five null features that would have a negative impact on model performance "furg_cnt," "burg_cnt," "flow_urg," "flow_cwr," and "flow_ece," we also eliminate four other almost null features like "std_idle," "bAvgBytesPerBulk," "bAvgPacketsPerBulk," and "bAvgBulkRate." After that, we dropped redundant instances and instances with missing values, and then the data was normalized before being split to 80% for training and 20% for testing using the hold-out validation strategy. In the FDL settings, 80% of the training data is again distributed to the participating clients. Figure 2.2 illustrates dataset class distribution after the preprocessing step using the t-SNE technique [20].

FDL-Based Model Training
We setup the FDL process using multiple clients holding local datasets. These local datasets were sampled from the main dataset and identically distributed according to the number of participating clients, all with the same feature vector as the main dataset.

Our proposed FDL-based training paradigm is demonstrated in Algorithm 1. Figure 2.3 depicts an organizational chart of our FDL-based Android malware detection method:

1. A coordinating central server starts the FDL process by initializing the global model architecture and corresponding global parameters

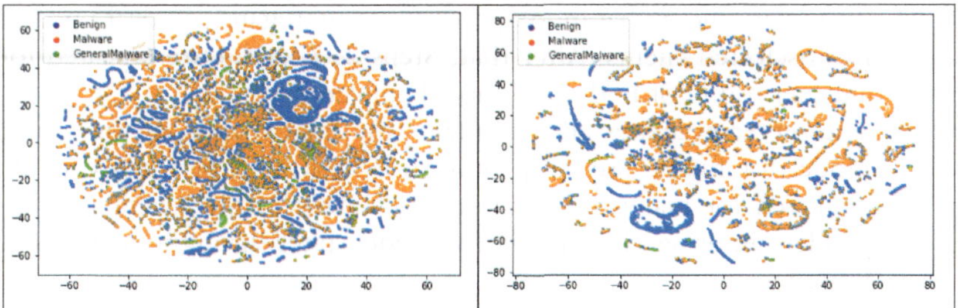

(a) Train data visualization (b) Test data visualization

Fig. 2.2 Exploring the high-dimensional AAGM2017-dataset using the t-SNE technique [20]. (**a**) Train data visualization. (**b**) Test data visualization

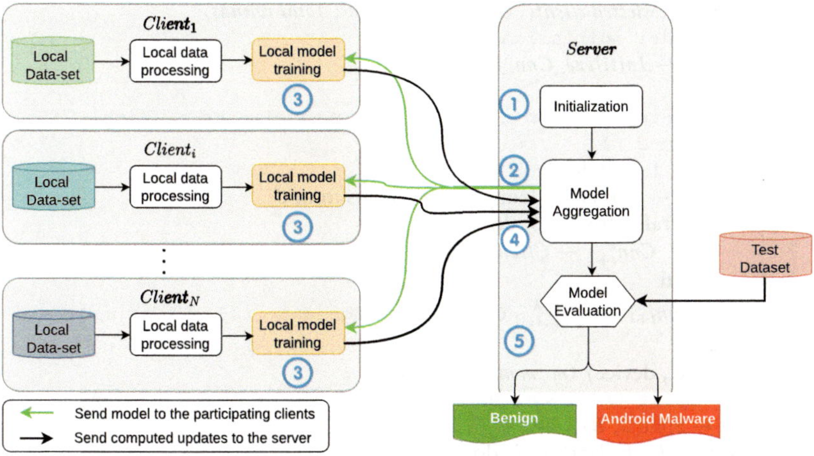

Fig. 2.3 A flow chart of the proposed FDL-based Android malware detection

such as the learning rate, the local batch size, and the local training epochs.

2. The server sends this information to pre-selected clients (i.e., clients with resource availability and sufficient training data) to compute local updates in an asynchronous manner.
3. Each client performs a number of local training epochs on the received model and then sends back the computed updates (i.e., the new model parameters) to the server.

4. To update the global model, the server aggregates all local updates from selected clients. After that, Steps 2, 3, and 4 are repeated for another round of FDL until model convergence.
5. The server evaluates and maintains the final version of the global model for future use. Depending on its local performance to be deployed for malware detection, each participating client is independent in preserving any global model states throughout the FDL training.

Algorithm 1: Federated learning based Android Malware detection [15]

Server (K :*Selected clients*, C :*Total clients*, R : *Total rounds*)
    `/* model initialization */`
    $Cnn_1 \leftarrow Initilize\_Cnn\_Model()$

    `/* Start FDL with a random selected clients at each round */`
    for $t = 1, .., R$ **do**
        $S_t \leftarrow$ Subset randomly chosen clients from C
        Parallel.for $k \in S_t$ **do**
            $Cnn^k_{t+1} \leftarrow Client(Cnn_t, k)$ `// Compute local updates`
        end
        $Cnn_{t+1} \leftarrow \frac{1}{K} \sum_{k=1}^{K} Cnn^k_{t+1}$ `// Aggregate all client updates`
    end

Client (i.e., device) (m :*model*, k :*client-Id*)
    `/*` `Split the local dataset` $\mathbb{D}$ `into` B `local data batch */`
    $\mathcal{B} \leftarrow$ Split($\mathbb{D}$, B)
    for i = *1,..,E* :*local epochs* **do**
        for $b \in \mathcal{B}$ **do**
            $m \leftarrow m - \eta \nabla f_c(m, b)$ `// Local client training`
        end
    end
    Send m to the Server

For the DL approach, different model architectures are well equipped and can be deployed to treat and handle the required malware behaviors, degrees of difficulty, and complexity. In our study, we selected the convolutional neural network model (CNN), a dedicated class of neural networks for data processing with a familiar network structure, designed mainly to extract discriminatory spatial features for model decisions. CNN networks are composed of a set of convolutional layers that use a mathematical operation called convolution and process perceptron layers [14].

2.4 Result and Discussion

The proposed FDL-based Android malware detection approach using CNN was tested in a free environment of the Google Colaboratory using the PyTorch library and GPU hardware accelerators. Table 2.2 depicts our experimental setting. Several experiments were carried out to adjust hyper-parameters and achieve an accurate and generalized detection model. The performance evaluation is conducted for malware detection (i.e., binary classification) using the following performance measures:

- *Accuracy (Acc)*: given by:

$$Acc = \frac{TP + TN}{TP + FP + TN + FN} \quad \text{where :} \tag{2.2}$$

TP: is the number of correctly classified positive samples.
TN: is the number of correctly classified negative samples.

Table 2.2 Experimental settings for federated learning

Subject	*Parameters*	*Values*
CNN Pytorch	Conv1d-1 Classifier	[1, 64, 70]
	Conv1d-2	[1, 32, 70]
	Conv1d-3	[1, 16, 70]
	Linear-4	[1, 32]
	Linear-5	[1, 2]
	Learning rate η	0.001
	Loss function	*CrossEntropyLoss*
	Activation function	*ReLu*
	Batch size	126
	Classification function	*SoftMax*
FDL	Clients Sets	[10, 20, 40]
	Data Distribution	IID
	Local epochs	[2,3]
	Total rounds	30
	Local Batch size	32

FP: is the number of wrongly classified positive samples
FN: is the number of wrongly classified negative samples.

- *Precision (Pr)*: the proportion of appropriate malware predictions (TP) to the total number of predicted malware outcomes, as given by :

$$Pr = \frac{TP}{TP + FP} \tag{2.3}$$

- *Detection rate (Dr)*: the proportion of appropriate malware predictions (TP) compared to the overall count of all samples that should have been detected as malware as given by :

$$Dr = \frac{TP}{TP + FN} \tag{2.4}$$

- *Time complexity*: the temporal complexity of the global model convergence and depends on the client's training time and model aggregation time. Considering that clients are training simultaneously, the time complexity would be the average time of all clients' training plus model aggregation. The temporal complexity of the server-client interaction was ignored.

Table 2.3 presents a comparison of detection performance with other comparable recent research using the AAGM-2017 dataset. The experimental settings differed, with different validation strategies and different test and training samples. For that, we implemented a centralized detection

Table 2.3 Comparison of performance between our proposed detection method and other related works using AAGM2017 dataset

Reference	*Classes*	*Acc*	*Pr*	*Recall*	*F1-score*	*Support*
Lashkari et al. 2018 [12]	Benign + Mal	0.91	0.91	N/A	N/A	N/A
Andresini et al. 2021 [3]	Benign	0.89	N/a	0.95	0.71	8000
	Malware			0.66		2000
Acharya et al. 2022 [1]	Benign			N/A		
	Malware	N/A	0.97	0.96	0.97	1915
Our centralized Cnn	Benign	0.84	0.87	0.89	0.88	41877
	Malware		0.78	0.76	0.77	22408
Our proposed FDL approach	Benign	0.837	0.85	0.91	0.88	41877
	Malware		0.80	0.71	0.75	22408

Acc: Accuracy, **Pr**: Precision, **Support** : Number of test instances

strategy using the same settings and compared it with the proposed FDL-based detection approach. The FDL-based model classified the "Benign" class, which represents normal apps, with a recall of 92%, and the "malware" class, which comprises all ten Android malware families, with a 71% of detection rate. The results demonstrate the efficiency of FDL, with practically the same performance as the centralized approach. However, these results of both detection approaches are not enough for real-world application, considering the high rate of false positives and false negatives as illustrated in Fig. 2.4.

Figure 2.5 illustrates a comparison of model accuracy, loss, and time complexity using different training approaches. In terms of time complexity, we can demonstrate the efficacy of the proposed FDL approach. However, when using a large number of participating clients, the global model's accuracy decreased from 83.74% to 78.47%, as depicted in Table 2.4.

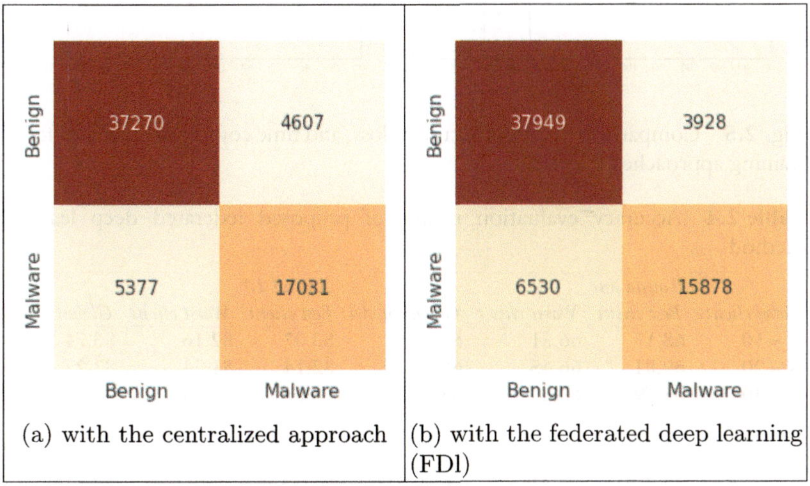

Fig. 2.4 Confusion matrix results. (**a**) with the centralized approach. (**b**) with the federated deep learning (FDl)

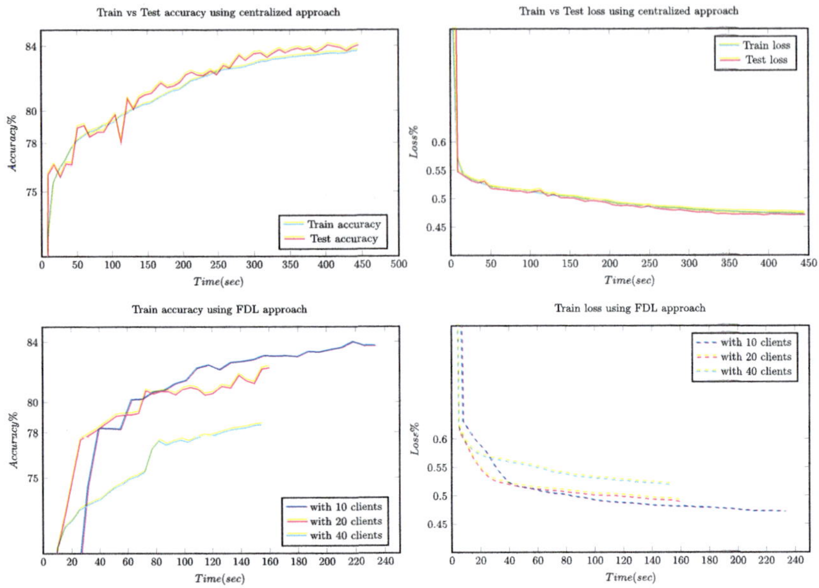

Fig. 2.5 Comparison of model accuracy, loss, and time complexity using different training approaches

Table 2.4 Accuracy evaluation results of proposed federated deep learning method

	Round one			*Round 10*		
Total clients	*Best client*	*Worst client*	*Global model*	*Best client*	*Worst client*	*Global model*
K = 10	68.17	66.31	60.95	83.07	82.16	83.74
K = 20	69.01	66.45	68.27	82.14	81.74	82.27
K = 40	65.79	63.6	65.34	78.05	76.38	78.47

2.5 Conclusion

In this chapter, we propose a novel, cost-effective DL-based Android malware system (FDL) leveraging the emergent federated learning paradigm. The analysis was conducted using the network layer features of malware samples to detect any variation from their normal behavior. Experimental results proved the efficiency and effectiveness of the proposed FDL paradigm compared with conventional centralized methods in terms of computation cost and privacy protection. However, the detection efficiency was not good enough when considering only network-based statistical features, and it was limited to only those sets of malware that require network connectivity and produce some abnormal network behavior. In the future, we intend to integrate local behavior with traffic behavior to efficiently detect large sets of malware. Also, we plan to improve the detection approach of our proposed FDL by employing non-identically distributed data as well as secure aggregation against emergent adversarial attacks.

References

1. Acharya S, Rawat U, Bhatnagar R (2022) A low computational cost method for mobile malware detection using transfer learning and familial classification using topic modelling. Appl Comput Intell Soft Comput 2022:1–22
2. Al-Fuqaha A, Guizani M, Mohammadi M, Aledhari M, Ayyash M (2015) Internet of things: a survey on enabling technologies, protocols, and applications. IEEE Commun Surv Tutorials 17(4):2347–2376
3. Andresini G, Appice A, Malerba D (2021) Autoencoder-based deep metric learning for network intrusion detection. Inf Sci 569:706–727
4. Arora A, Garg S, Peddoju SK (2014) Malware detection using network traffic analysis in android based mobile devices. In: 2014 eighth international conference on next generation mobile apps, services and technologies. IEEE, New York, pp 66–71
5. Aslan ÖA, Samet R (2020) A comprehensive review on malware detection approaches. IEEE Access 8:6249–6271
6. Garg S, Peddoju SK, Sarje AK (2017) Network-based detection of android malicious apps. Int J Inf Secur 16(4):385–400
7. Hamouda D, Ferrag MA, Benhamida N, Seridi H (2021) Intrusion detection systems for industrial internet of things: A survey. In: 2021 International Conference on Theoretical and Applicative Aspects of Computer Science (ICTAACS). IEEE, New York, pp 1–8

8. Hamouda D, Ferrag MA, Benhamida N, Seridi H (2022) PPSS: a privacy-preserving secure framework using blockchain-enabled federated deep learning for industrial IoTs. Pervasive Mob Comput 88:101738
9. Iadarola G, Martinelli F, Mercaldo F, Santone A (2021) Towards an interpretable deep learning model for mobile malware detection and family identification. Comput Secur 105:102198
10. Karbab EB, Debbabi M, Derhab A, Mouheb D (2018) Maldozer: automatic framework for android malware detection using deep learning. Digit Investig 24:S48–S59
11. Kim T, Kang B, Rho M, Sezer S, Im EG (2018) A multimodal deep learning method for android malware detection using various features. IEEE Trans Inf Forensics Secur 14(3):773–788
12. Lashkari AH, Kadir AFA, Gonzalez H, Mbah KF, Ghorbani AA (2017) Towards a network-based framework for android malware detection and characterization. In: 2017 15th annual conference on privacy, security and trust (PST). IEEE, New York, pp 233–23309
13. Lashkari AH, Kadir AFA, Taheri L, Ghorbani AA (2018) Toward developing a systematic approach to generate benchmark android malware datasets and classification. In: 2018 International Carnahan Conference on Security Technology (ICCST). IEEE, New York, pp 1–7
14. LeCun Y, Bengio Y, Hinton G (2015) Deep learning. Nature 521(7553):436–444
15. McMahan B, Moore E, Ramage D, Hampson S, y Arcas BA (2017) Communication-efficient learning of deep networks from decentralized data. In: Artificial intelligence and statistics. PMLR, New York, pp 1273–1282
16. Qamar A, Karim A, Chang V (2019) Mobile malware attacks: review, taxonomy and future directions. Futur Gener Comput Syst 97:887–909
17. Rey V, Sánchez PMS, Celdrán AH, Bovet G (2022) Federated learning for malware detection in IoT devices. Comput Netw 204:108693
18. Souri A, Hosseini R (2018) A state-of-the-art survey of malware detection approaches using data mining techniques. Hum-centric Comput Inf Sci 8(1):1–22

19. Tam K, Feizollah A, Anuar NB, Salleh R, Cavallaro L (2017) The evolution of android malware and android analysis techniques. ACM Comput Surv (CSUR) 49(4):1–41
20. Wattenberg M, Viégas F, Johnson I (2016) How to use t-SNE effectively. Distill **1**(10):e2
21. Yuan Z, Lu Y, Xue Y (2016) Droiddetector: android malware characterization and detection using deep learning. Tsinghua Sci Technol 21(1):114–123
22. Zhu D, Jin H, Yang Y, Wu D, Chen W (2017) Deepflow: deep learning-based malware detection by mining android application for abnormal usage of sensitive data. In: 2017 IEEE symposium on computers and communications (ISCC). IEEE, New York, pp 438–443
23. Zulkifli A, Hamid IRA, Shah WM, Abdullah Z (2018) Android malware detection based on network traffic using decision tree algorithm. In: International conference on soft computing and data mining. Springer, Berlin, pp 485–494

CHAPTER 3

ASParseV3: Auto-Static Parser and Customizable Visualizer

Iman Almomani, Rahaf Alkhadra, and Mohanned Ahmed

3.1 Introduction

Our modern world is rapidly moving toward digitalization and automation, where everything is converging into an automated version. As technology takes over our lives, we are at the start of the 4th industrial revolution, which mainly focuses on a world that relies heavily on technology and innovation. The use of technology not only provides us with convenience but comfort as well. However, the rapid development of technology comes at the price of ensuring cybersecurity. Attackers are finding many ways to achieve their malicious goals, which requires us to take precautions to

I. Almomani (✉)
Security Engineering Lab, Prince Sultan University, Riyadh, Saudi Arabia

Computer Science Department, The University of Jordan, Amman, Jordan
e-mail: imomani@psu.edu.sa; i.momani@ju.edu.jo

R. Alkhadra • M. Ahmed
Security Engineering Lab, Computer Science Department, Prince Sultan University, Riyadh, Saudi Arabia
e-mail: rkhadra@psu.edu.sa; mqasem@psu.edu.sa

I. Almomani et al. (eds.), *Cyber Malware*, Security Informatics and Law Enforcement, https://doi.org/10.1007/978-3-031-34969-0_3

face such security issues. One of the most popular and common forms of security invasion in our digital world is using malicious code, often referred to as malware [27]. Malware is a code written by security attackers to intrude into a specific computer system or software to perform malicious acts such as stealing data or causing damage. For example, malware could be in different forms, such as worms, viruses, trojans, spyware, adware, or ransomware. Therefore, it is essential to protect any system from malware. This can be done by detecting the malware and then classifying which type it is. A tremendous amount of research has been conducted in the past years regarding the topic of malware detection and classification [11].

According to recent reports, malware generation and creation have been increasing rapidly on a daily basis. It is estimated that around one million malware files are created daily [31]. This increase could seriously threaten the economy, both financially and technically. The increase in cyber threats and crimes costs the economy around 1 trillion dollars in 2022 for cyber insurance, which results in an increase of 50% in comparison to the past 2 years [12]. The term malware refers to any malicious entity that changes the original behavior by utilizing software flaws and vulnerabilities. In this chapter, the term malware will be used to refer to any malicious software that may include any of the following malware families, ransomware, adware, viruses, or keyloggers [11].

Depending on the purpose and behavior of the malware, it is categorized into different families. Every family has common features. For instance, stealing information, creating vulnerability, and denial of service are all examples of malware behavior. Such behaviors are essential in detecting malware since this information will be used to analyze the software and categorize it into benign or malware [35]. To differentiate between malicious and benign apps, we need to scan the program code first, extract its features, and analyze them [6]. Features extraction can be achieved through two main ways: static analysis [3] and dynamic analysis [13]. Another possible way is to use hybrid analysis [2], a combination of the previous two [25]. Static analysis is concerned with contextual data from the source code without running the program. However, dynamic analysis involves executing the program and extracting the runtime features. The hybrid analysis uses both contextual and runtime features to detect malware [11].

Over the years, researchers have been developing new techniques for malware detection. The latest trend in this field is using machine learning for malware detection. However, this technique cannot be used without analyzing the program code and extracting important features that help in discriminating the malware families [22]. It is possible to evade the risk of malware if the related features are available. Therefore, a collection of advanced detection methods using machine learning depends on feature engineering as well as reverse engineering [33]. Feature engineering is a technique used to manipulate unstructured data into features that can be understandable by the computer or machine [32]. However, other techniques, such as binary obfuscation, can be used by attackers to design a reverse engineering resistant file [30]. Moreover, deep learning can be used in an advanced model of neural networks to capture features, learn, and adapt during training. Even though a few studies report the use of deep learning, some do not discuss the scalability and different architectures enough for malware detection [5, 33].

One of the main benefits of using static analysis over any other technique is that this analysis does not require executing the program, making it a safer choice to apply [25]. Moreover, another vital benefit is examining the code without regard to the diversity of IoT architecture or the physical capabilities of an IoT device. Hence, the analysis considers all possible inspection methods with no reference to the physical performance [24]. Furthermore, due to the nature of the static analysis, the malware may not be able to avoid, hide, and/or obfuscate during the analysis process because it runs passively [34]. Finally, its automation characteristic is what makes static analysis prominent and outstanding [16].

Therefore, this chapter introduces a new comprehensive static parsing software called ASParseV3. It is an extension to ASParseV1 [1]. It is a GUI-based tool with various features such as **(a)** selecting many files or directories to be scanned in one experiment, **(b)** adding or removing keywords/features, **(c)** filtering the keywords/features and specific file types, **(d)** efficient scanning process as many files are scanned simultaneously, **(e)** providing customizable visualization dashboards with the ability to export the chart(s), and **(f)** exporting the results in different formats such as JSON and CSV.

The rest of the chapter sections present and discuss the related works regarding malware analysis techniques, malware detection, and the use of static analysis for malware detection. Moreover, they present the proposed developed software (ASParseV3), which performs static features extraction and parsing. Also, the chapter demonstrates a use case of Android OS malware static features extraction using the ASParseV3 software. Finally, conclusions with a summary of possible future works are presented.

3.2 Related Works

Parsing the features of source code is potentially utilized in estimating the software performance, reverse engineering, and static analysis [20]. However, the extracted features can be represented in different formats such as gray-scale images, structural entropy, or JSON file [15]. Moreover, the extracted features can be further deployed in various fields. For instance, the authors of [21] have developed a tool named DeepTLS to analyze encrypted traffic by extracting the features from the network packets. In [28], the python-Evtx-parser (pexp) has been developed to parse the required features to detect Lateral Movement Attacks. In a nutshell, Table 3.1 demonstrates a comparison among related works.

Several tools have been proposed to perform static parsing in Android platform [1, 8, 23]. Khalid et. al. proposed a memory parsing tool for Android applications [19]. The authors of [17] have developed Sena TLS-Parser, a tool that automates the software testing process by parsing the Android source code. Initially, the Android source code is imported into the Eclipse environment. Subsequently, Sena TLS-Parser scans the code and generates the required test cases. Another approach that utilizes static parsing in enhancing the development of Android applications is by recommending a suitable API for the Android developer based on the parsing results. In [36], the authors have developed APIMatchmaker, a tool that recommends the best API usage by parsing similar Android apps.

Parsing Android source code can further be deployed in detecting malicious applications. In [26], the authors have parsed the suspect methods of two Android apps in order to extract their similarities using their proposed tool, StrAndroid. Consequently, they identified the potential malicious behaviors that are shared between the two apps. Additionally, Android permissions can be parsed in order to rank the risk of the malicious application. Dharmalingam et al. proposed a permission grading scheme that extracts and defines the required permissions in an Android app and rates the risk of the app accordingly [14]. In their proposed scheme, the Manifest file is parsed to extract the defined permission in the app. Subsequently, the extracted permissions are fed into the feature encoder to be further utilized in the deep neural algorithm for detecting malware applications. However, static analysis can be combined with dynamic analysis to increase the efficiency of malware detection. In [2], the authors have applied static analysis as a prior stage to implementing the dynamic analysis.

Table 3.1 A comparison among existing parsing tools

Work	*Year*	*Tool name*	*Aim*	*Scanned File Type*	*Has GUI?*	*Customization*	*Number of Extracted Features*	*Exported File Format*	*Scanning Environment*	*Is a System?*	*Developing Language*
[21]	2022	DeepTLS	To analyze encrypted traffic by extracting the features from the network packets	Pcap	Yes	No	70+	JSON	NM	Yes	C++
[15]	2022	PE Parser	To parse executable binary files in order to extract required features for malware detection	Exe files	No	No	14	Gray-scale images, structural entropy.	NM	No	Python
[28]	2022	python-Evtx-parser (pexp)	To parse the required features from the network packet to detect Lateral Movement Attacks	.evtx files	No	No	NM	xml objects	Windows 10, macOS, Ubuntu	No	Python
[17]	2022	TLS-Parser	To automate the software testing process by parsing the Android source code	.java	No	Yes	NM	.java	Windows	No	Java
[26]	2020	StrAndroid,	To parse the suspect methods of two Android apps in order to extract their similarities	APK	No	No	NM	Text file	NM	Yes	Python
[36]	2022	API Matchmaker	To recommend the best API usage by parsing similar Android apps	APK	No	No	NM	Text file	NM	No	Java
[18]	2021	PetaDroid	To cluster the malware families based on static analysis for Android OS	APK	No	No	300+	Text file	NM	Yes	Python, Bash
[14]	2020	Permission Grader	To grade the risk level of Android malware app based on its extracted permissions	Manifest file	No	Yes	1000	Text file	NM	Yes	NM
[29]	2020	DroidPortrait	To utilize the extracted Android permissions and API calls in developing a malware portrait	Manifest file, class.dex	No	Yes	50,000	PNG	NM	Yes	NM
This work	2022	ASParseV3	To propose a GUI-based, customizable, and comprehensive static parsing tool with the ability to export results/charts in different formats	Flexible (Any)	Yes	Yes	Unlimited	CSV Meta-Data: Json Graph: PNG	Cross-platform	Yes	Python

The efficiency of the parsing approach highly affects the overall static analysis process. The authors of [18] applied canonical representation to enhance the parsing process for Android code by developing the static analyzing tool, PetaDroid. The core of this proposed solution is to define the application's behavior by tracking the used APIs and the app's actions. Consequently, fingerprinting the malware applications. Besides the API calls, the permissions can be utilized to determine the malicious application's behavior. In [29], the APK file has been decomposed using APKtool to retrieve the Manifest file and class.dex file. The aforementioned files were parsed to extract the permissions and the API calls, respectively. Then, multidimensional behavior analysis was conducted on the extracted features to develop a malware portrait. Even though there are many static parsing tools, they are not flexible in accepting many file systems and can extract only a limited number of features. Moreover, they do not have a customizable graphical user interface (GUI). Therefore, there is a need for a customizable GUI-based system with the ability to scan an unlimited number of features on various file systems.

3.3 Proposed System

There is a need for user-friendly, extensible, and flexible software. This chapter introduces the third version of the Android Static Parse (ASParse). The tool ASParse-V3 is an improvement to the previous versions. It is a cross-platform, portable, and general tool that performs static analysis and features parsing for any file type while supporting different operating systems. This version of ASParse is efficient and fast due to its concurrent scanning characteristic. Furthermore, ASParseV3 can be used as a preprocessing method for static feature extraction to construct datasets for subsequent processing through ML/DL models due to its feature of exporting the results to JSON and CSV files. For instance, the previous versions of the ASParse tool were used to extract static features and develop different types of datasets [1]. For example, [4, 7] utilized the ASParse tool to extract the API and permissions of thousands of Android applications. The extracted features created a dataset that helped detect Ransomware apps with high accuracy.

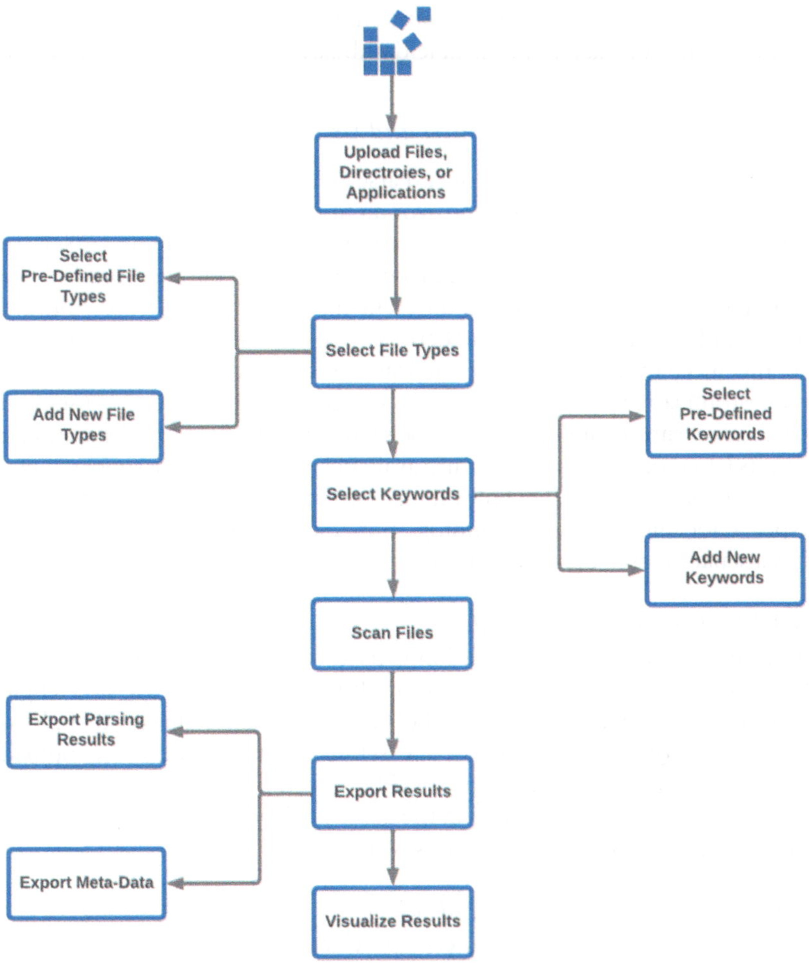

Fig. 3.1 Flow structure of the proposed system

3.3.1 System Overview

To illustrate the system flow, Fig. 3.1 shows how the ASParseV3 application generally works. The first step is uploading the files, directories, or multiple directories. The second step is choosing a set of predefined features or adding specific features. Then, moving to the third step, the system scans the files to export the results. Finally, after the results are exported, they can be visualized via a customizable dashboard.

3.3.2 Features and User Interfaces

The scalability and portability of ASParseV3 are achieved by integrating it with a portable development environment that also makes the software cross-platform to be installed on various operating systems (OSs). In addition, the software's scope can be used as general and specific. For example, it can scan and parse different input formats, such as Android and Windows applications. Furthermore, ASParseV3 is user-friendly due to the modern graphical user interface (GUI) that is easy to use and its customizability based on the user's needs. For instance, the user can customize features and file types to be scanned and customize the scanning results based on the filtering feature available on the results dashboard. The system process is divided mainly into five steps: uploading files, selecting file types, choosing keywords, scanning, and results visualization. Each phase has a separate user-friendly window.

3.3.2.1 Uploading Files Window

The first window of the application is used to upload files or applications to be scanned. The user can upload multiple files, directories, or a single directory. As Fig. 3.2 illustrates, the button "Add" is clicked to upload the applications, which opens a file selector dialog window to upload files/directories. All uploaded files will be shown on a panel field. The user may also clear the uploaded files in the panel field by clicking on the "Clear" button and adding new applications when needed.

3.3.2.2 Selecting File Types Window

The second window allows users to select files of specific types (file extensions) to be scanned. Figure 3.3a shows a sample of Android OS file types. The user may choose one or multiple types by checking the checkbox. Moreover, the user can customize the file types by adding or deleting types by clicking on the settings icon on the top right of Fig. 3.3a. The settings button opens a new window for editing, as Fig. 3.3b illustrates. The user can write the file types in the text field and then click on the button "Add" to add them to the current panel. The user can also delete any newly defined types by clicking on the button "Remove." By default,

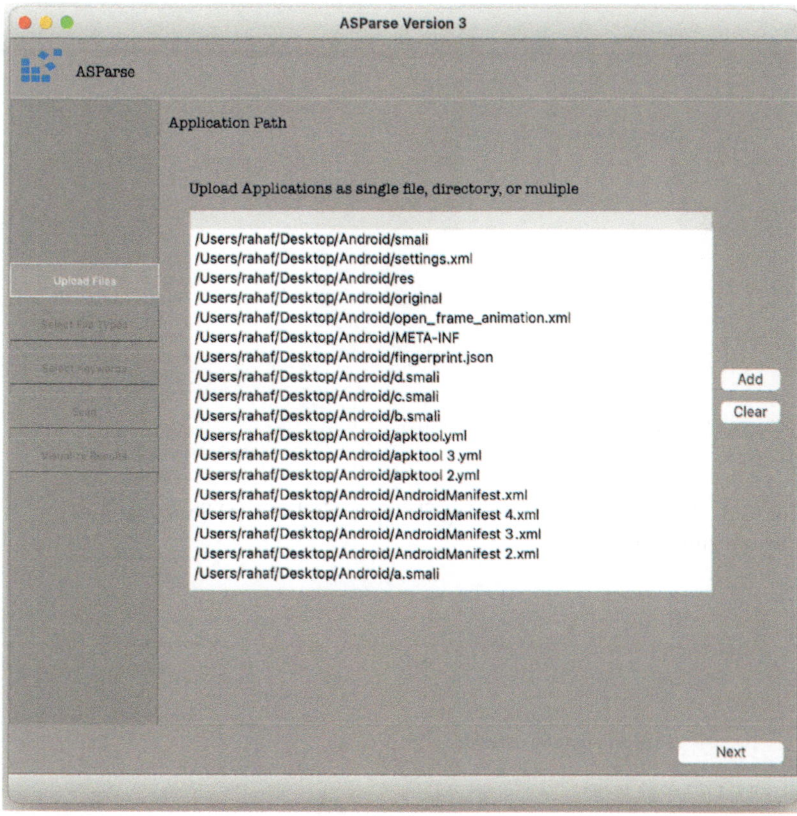

Fig. 3.2 Uploading applications window

if no checkboxes were chosen, all predefined file types will be included in the scanning process.

3.3.2.3 Selecting Keywords Window

The third window allows users to select the keywords to look for while scanning. Figure 3.3a shows a sample of Android OS file types. However, the user can customize the features through the settings window by adding or deleting keywords by clicking on the settings icon on the top right of the window (as shown in Fig. 3.4a). Similar to the file types editing feature,

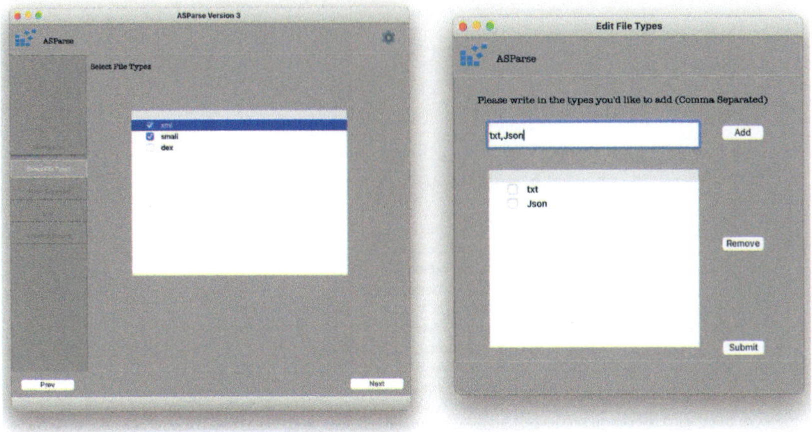

(a) Selecting Window.

(b) Customizing Window.

Fig. 3.3 Selecting and customizing file types windows. (**a**) Selecting Window. (**b**) Customizing Window

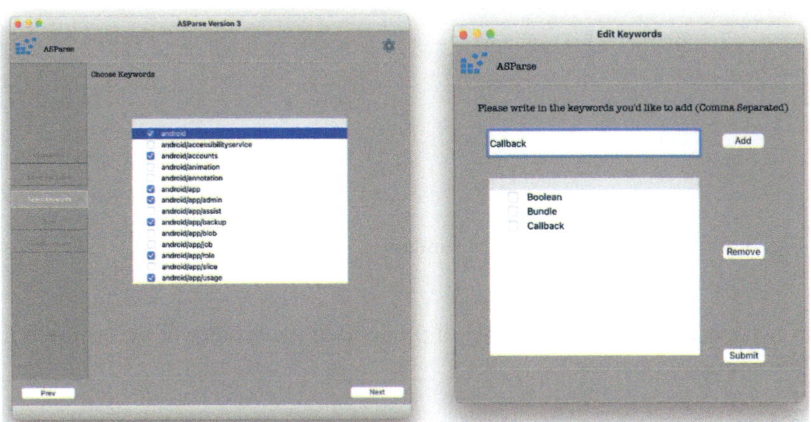

(a) Selecting Window.

(b) Customizing Window.

Fig. 3.4 Selecting and customizing keywords windows. (**a**) Selecting Window. (**b**) Customizing Window

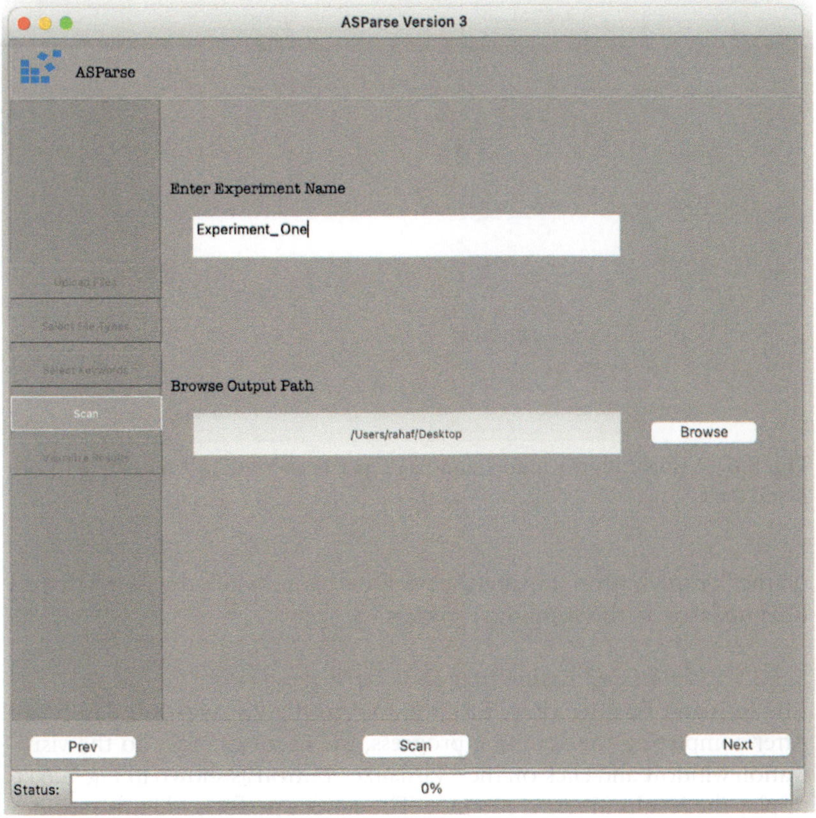

Fig. 3.5 Scanning window

the settings button can be used to edit the list of keywords, as illustrated in Fig. 3.4b.

3.3.2.4 Scanning Window

The fourth window allows users to add the configuration values of an experiment, such as the experiment name and the path used to save the results, as shown in Fig. 3.5. Then, the scanning process begins by clicking

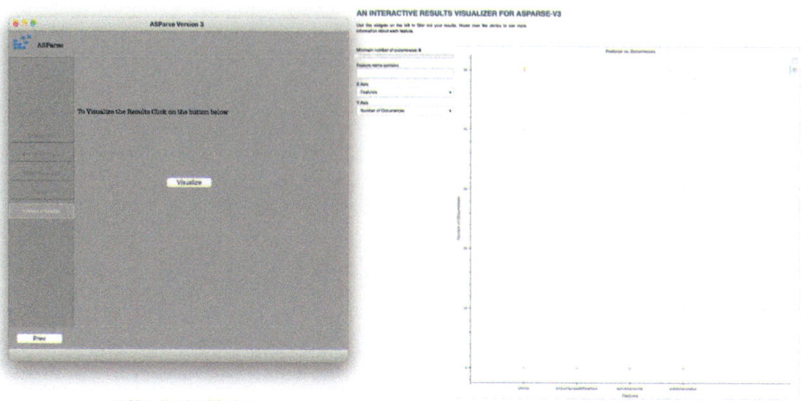

(a) Visualization Window.

(b) Dashboard Page.

Fig. 3.6 Visualization window and page. (**a**) Visualization Window. (**b**) Dashboard Page

on the "Scan" button. Finally, the progress bar provides the user with real-time updates on the scanning progress.

3.3.2.5 *Visualizing Results and Dashboard Window*

The fifth and final window links the tool to the visualization dashboard. After completing the scanning progress, the user can move to the visualization window and click on the "Visualize" button as shown in Fig. 3.6a to display the results in terms of a plot. The actions performed in this window do not affect the scanning results. It is a complimentary step for results visualization and filtering. However, this step cannot be completed without performing the scanning. When visualization is activated, a dashboard page opens in the browser. The dashboard is where the user can visualize the parsing results. The plot's X-axis represents the features (keywords), and the Y-axis represents the number of occurrences. As Fig. 3.6b illustrates, the dashboard is customizable based on the user's preference. For instance, the user may filter out and visualize the results according to the minimum number of feature occurrences and features containing a specific string or substring. Also, the resulting graph (plot) can be exported as an image using the saving button on the right of the plot. This can help the researchers/experts to share their results conveniently.

3.3.3 *Use Case*

To demonstrate the tool, Android benign samples and malware samples were used. The samples come in the form of an Android Package Kit (APK). The APKs contain all software details, including source code, permissions, and APIs used. However, APKs are compressed files that need reverse engineering to recover the application code [9]. APKTool[1]s was used to decompile the apps and extract the source files. Afterward, the decompiled APKs were fed to the ASParse tool.

3.3.3.1 Data Collection

For data collection, two sources were used, Drebin Dataset [10] and APKCombo.[2] The Drebin Dataset contained 5560 malware samples belonging to 179 malware families. On the other hand, the benign data samples were downloaded through APKCombo. Ten samples were randomly chosen from the Drebin dataset, along with ten samples from APKCombo. To ensure that the apps downloaded from APKCombo are benign, they were scanned by a well-known website called VirusTotal.[3] This website offers tens of Antivirus engines that are specialized in detecting different types of malware.

3.3.3.2 Tests and Results

The experiment was performed on a sample of 10 benign APKs and 10 malicious samples from the Derbin dataset. First, all files were added to the application upload field. Then, all predefined file types were chosen. Afterward, six keywords from the predefined ones were chosen, including android, android/animation, and android/app. In addition to the keywords Bundle and Button and Callback. After clicking on the visualization button in the final window, the application will shift to the dashboard, where the plot will be displayed with the ability to save the plot after customizing it. Figure 3.7 illustrates the saved plot sample. Moreover, Fig. 3.8 illustrates a sample of the saved plot where it illustrates the details of each data point on the plot. Furthermore, Table 3.2 demonstrates a sample of the resulting CSV. Finally, Fig. 3.9 represents the JSON metadata file resulting from the scan.

---

[1] https://ibotpeaches.github.io/Apktool/.

[2] https://apkcombo.com/.

[3] https://www.virustotal.com/gui/home/upload.

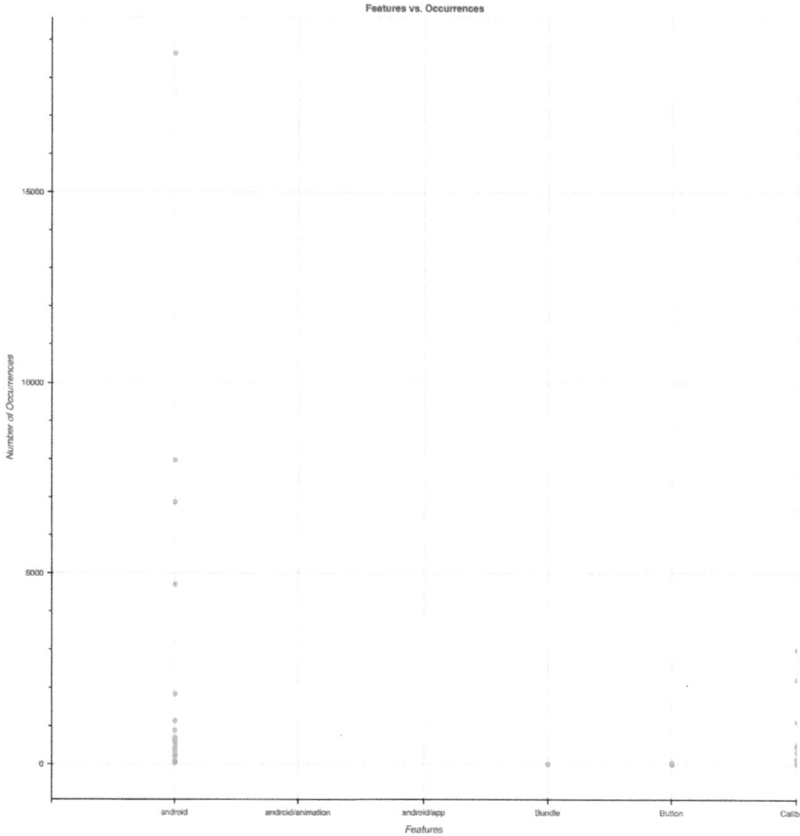

Fig. 3.7 Features vs. Occurrences Plot

3.3.3.3 Validation

The validation process for ASParseV3 was carried out thoroughly to ensure that its performance, user interface (UI), and user experience (UX) met the required needs. The Security Engineering Lab (SEL) conducted the validation and compared the scanning results of ASParseV3 with previous releases of ASParse. In addition, VirusTotal was used to retrieve information such as permissions used in the applications/APKs to compare with ASParseV3 and verify further its scanning results' accuracy. To validate the use case, VirusTotal was used to collect the permissions used by the APK.

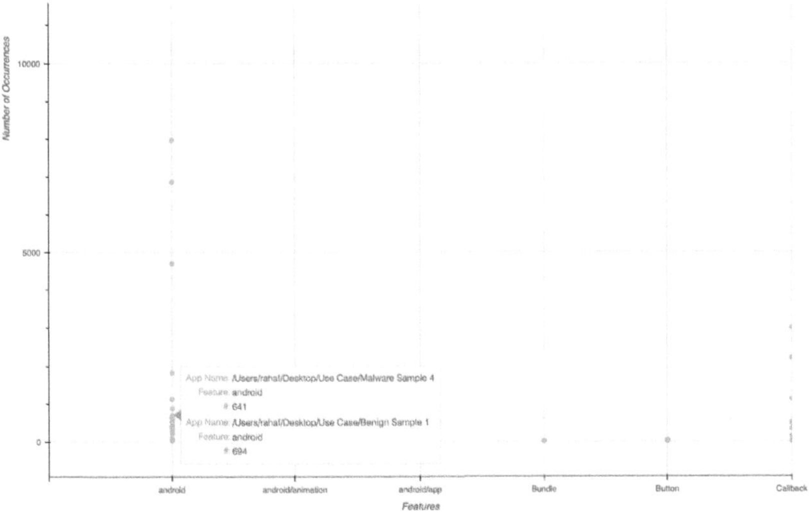

Fig. 3.8 Data point details

Figure 3.10 shows a sample of the permissions used by the APK validation test sample. The resulting permissions were then used to scan the same APK using ASParseV3. The results showed that ASParseV3could scan the uploaded APK and accurately report the number of occurrences for each permission. Overall, the validation process demonstrates that ASParseV3 is a reliable and efficient tool for scanning applications and APKs features such as permissions. The comparison with previous releases and the use of VirusTotal helped ensure the scanning results' accuracy. For example, Table 3.3 illustrates the number of occurrences of each permission found by ASParseV3 during the validation process. Moreover, using ASParseV3 to scan the same application without specifying any keywords has resulted in showing additional permissions/API calls other than the ones retrieved from VirusTotal as Table 3.4 illustrates. Hence, this validates the accuracy of the ASParseV3 and its additional capabilities compared with similar tools.

Table 3.2 The resulting CSV from the use case

fileName	*Android*	*Android/ animation*	*Android/app*	*Button*	*Bundle*	*Callback*
/Users/rahaf/Desktop/Use Case/Malware Sample 5	212	0	0	10	0	0
/Users/rahaf/Desktop/Use Case/Malware Sample 7	221	0	0	4	0	0
/Users/rahaf/Desktop/Use Case/Malware Sample 8	53	0	0	0	0	0
/Users/rahaf/Desktop/Use Case/Malware Sample 4	641	0	0	15	1	9
/Users/rahaf/Desktop/Use Case/Malware Sample 1	1834	0	0	0	1	0
/Users/rahaf/Desktop/Use Case/Malware Sample 10	355	0	0	0	0	0
/Users/rahaf/Desktop/Use Case/Malware Sample 3	535	0	0	9	0	0
/Users/rahaf/Desktop/Use Case/Malware Sample 6	254	0	0	2	1	5
/Users/rahaf/Desktop/Use Case/Benign Sample 5	37	0	0	0	0	0
/Users/rahaf/Desktop/Use Case/Malware Sample 2	96	0	0	0	1	0
/Users/rahaf/Desktop/Use Case/Benign Sample 3	0	0	0	0	0	0
/Users/rahaf/Desktop/Use Case/Malware Sample 9	430	0	0	9	0	18
/Users/rahaf/Desktop/Use Case/Benign Sample 4	596	0	0	4	1	36
/Users/rahaf/Desktop/Use Case/Benign Sample 1	694	0	0	23	4	148
/Users/rahaf/Desktop/Use Case/Benign Sample 7	880	0	0	5	5	471
/Users/rahaf/Desktop/Use Case/Benign Sample 2	1128	0	0	13	7	1121
/Users/rahaf/Desktop/Use Case/Benign Sample 10	4709	0	0	21	7	333
/Users/rahaf/Desktop/Use Case/Benign Sample 9	6871	0	0	10	3	517
/Users/rahaf/Desktop/Use Case/Benign Sample 6	7968	0	0	38	15	2215
/Users/rahaf/Desktop/Use Case/Benign Sample 8	18638	0	0	40	16	3010

```
{
  {
    "ApplicationPath": [
        "/Users/rahaf/Desktop/Use Case/Malware Sample 5",
        "/Users/rahaf/Desktop/Use Case/Malware Sample 4",
        "/Users/rahaf/Desktop/Use Case/Malware Sample 3",
        "/Users/rahaf/Desktop/Use Case/Malware Sample 2",
        "/Users/rahaf/Desktop/Use Case/Malware Sample 1",
        ...
        "/Users/rahaf/Desktop/Use Case/Benign Sample 5",
        "/Users/rahaf/Desktop/Use Case/Benign Sample 4",
        "/Users/rahaf/Desktop/Use Case/Benign Sample 3",
        "/Users/rahaf/Desktop/Use Case/Benign Sample 2",
        "/Users/rahaf/Desktop/Use Case/Benign Sample 1"
    ],
    "OutputPath": "/Users/rahaf/Desktop",
    "filetypes": [
        "xml",
        "smali",
        "dex"
    ],
    "selectedFileTypes": [
        "smali",
        "xml",
        "dex"
    ],
    "keywords": [
        "android",
      "android/accessibilityservice",
        "android/accounts",
        "android/animation",
        "android/annotation",
        "android/app",
        "android/app/admin",
        "android/app/assist",
        "android/app/backup",
        "android/app/blob",
        "android/app/job",
        "android/app/role",
        "android/app/slice",
        "android/app/usage",
        "android/appwidget",
        "android/bluetooth",
        "Button",
        "Bundle",
        "Callback"
        ...

    ],
    "selectedKeywords": [
        "android",
        "android/animation",
        "android/app",
        "Button",
        "Bundle",
        "Callback"
    ],
    "ExperimentName": "Experiment_One"
}

}
```

Fig. 3.9 Metadata JSON content for the use case

Permissions

- ⓘ android.permission.RECEIVE_BOOT_COMPLETED
- ⓘ android.permission.ACCESS_WIFI_STATE
- ⓘ com.google.android.gms.permission.AD_ID
- ⓘ com.google.android.finsky.permission.BIND_GET_INSTALL_REFERRER_SERVICE
- ⓘ com.android.vending.BILLING

Fig. 3.10 APK permissions from VirusTotal

Table 3.3 Validation results

/Users/rahaf/Desktop/PSU/Use Case/Benign Sample 1	
Permissions	Occurrences
android.permission.RECEIVE_BOOT_COMPLETED	1
android.permission.ACCESS_WIFI_STATE	4
com.google.android.gms.permission.AD_ID	3
com.google.android.finsky.permission.BIND_GET_INSTALL_REFERRER_SERVICE	2
com.android.vending.BILLING	1

Table 3.4 ASParseV3 additional permissions and calls

/Users/rahaf/Desktop/PSU/Use Case/Benign Sample 1	
Permissions and calls	Occurrences
Android	18474
CallbackHandler	117
CameraAccessException	14
Certificate	285
Connection	1522
CookieSyncManager	1
DownloadRequest	8
FragmentHostCallback	3
LruCache	2
INTERNET	25

3.4 Conclusion and Future Work

This chapter proposed a third version of ASParse software as a parsing and static analysis tool. The analysis results can be used to feed machine learning algorithms and deep learning models for malware analysis and detection. Moreover, a demonstration was presented on Android OS applications showing the system's capabilities. In future work, the ASParse tool will be used to carry on with malware detection using ML and DL algorithms and models. Moreover, it will be enhanced in terms of performance and user experience.

Acknowledgments The authors would like to thank the support of Prince Sultan University. Moreover, this research was done during the author Iman Almomani's sabbatical year 2021/2022 from the University of Jordan, Amman—Jordan.

References

1. Al Khayer A, Almomani I, Elkawlak K (2020) ASAF: android static analysis framework. In: 2020 first international conference of smart systems and emerging technologies (SMARTTECH). IEEE, New York, pp 197–202
2. Almohaini R, Almomani I, AlKhayer A (2021) Hybrid-based analysis impact on ransomware detection for Android systems. Appl Sci 11(22):10976
3. Almomani I, Ahmed M, El-Shafai W (2022) Android malware analysis in a nutshell. PloS One 17(7):e0270647
4. Almomani I, AlKhayer A, Ahmed M (2021) An efficient machine learning-based approach for Android v. 11 ransomware detection. In: 2021 1st international conference on artificial intelligence and data analytics (CAIDA). IEEE, New York, pp 240–244
5. Almomani I, Alkhayer A, El-Shafai W (2022) An automated vision-based deep learning model for efficient detection of android malware attacks. IEEE Access 10:2700–2720
6. Almomani I, Khayer A (2019) Android applications scanning: the guide. In: 2019 International conference on computer and information sciences (ICCIS). IEEE, New York, pp 1–5
7. Alsoghyer S, Almomani I (2019) Ransomware detection system for Android applications. Electronics 8(8):868
8. Anupama ML, et al (2021) Detection and robustness evaluation of android malware classifiers. J Comput Virol Hacking Tech 18(3):1–24
9. Ardito L, et al (2020) Automated test selection for Android apps based on APK and activity classification. IEEE Access 8:187648–187670

10. Arp D, et al (2014) Drebin: effective and explainable detection of android malware in your pocket. In: NDSS, vol. 14, pp 23–26
11. Aslan ÖA, Samet R (2020) A comprehensive review on malware detection approaches. IEEE Access 8:6249–6271
12. Cremer F, et al (2022) Cyber risk and cybersecurity: a systematic review of data availability. In: The Geneva Papers on Risk and Insurance-Issues and Practice, pp 1–39
13. Dai Y, et al (2019) SMASH: a malware detection method based on multifeature ensemble learning. IEEE Access 7:112588–112597
14. Dharmalingam VP, Palanisamy V (2021) A novel permission ranking system for android malware detection—the permission grader. J Ambient Intell Humaniz Comput 12(5):5071–5081
15. Gibert D (2022) PE Parser: A Python package for Portable Executable files processing. Software Impacts 13:100365
16. Gosain A, Sharma G (2015) Static analysis: a survey of techniques and tools. In: Intelligent computing and applications. Springer, Berlin, pp 581–591
17. Ibrahim R, et al (2022) Sena TLS-Parser: a software testing tool for generating test cases. Int J Adv Comput Sci Appl 13(6):397–403
18. Karbab EB, Debbabi M (2021) Resilient and adaptive framework for large scale android malware fingerprinting using deep learning and NLP techniques. arXiv e-prints arXiv–2105
19. Khalid Z, et al (2022) Forensic investigation of Cisco WebEx desktop client, web, and Android smartphone applications. Ann Telecommun 78:1–26
20. Laaber C, Basmaci M, Salza P (2021) Predicting unstable software benchmarks using static source code features. Empir Softw Eng 26(6):1–53
21. Liu Z (2022) DeepTLS: comprehensive and high-performance feature extraction for encrypted traffic. arXiv preprint arXiv:2208.03862
22. Lu T, et al (2020) Android malware detection based on a hybrid deep learning model. Secur Commun Netw 2020:1–11
23. Mahr A, et al 2022 Auto-Parser: Android Auto and Apple CarPlay Forensics. In: International Conference on Digital Forensics and Cyber Crime. Springer, Berlin, pp 52–71

24. Ngo Q-D, et al (2020) A survey of IoT malware and detection methods based on static features. ICT Express 6(4):280–286
25. Omer MA, et al (2021) Efficiency of malware detection in android system: a survey. Asian J Res Comput Sci 7(4):59–69
26. Pasetto M, Marastoni N, Preda MD (2020) Revealing similarities in android malware by dissecting their methods. In: 2020 IEEE European Symposium on Security and Privacy Workshops (EuroS&PW). IEEE, New York, pp 625–634
27. Shukla S (2022) Design of secure and robust cognitive system for malware detection. arXiv preprint arXiv:2208.02310
28. Smiliotopoulos C (2022) Use of Sysmon tool to detect lateral movement attacks
29. Su X, et al (2020) DroidPortrait: android malware portrait construction based on multidimensional behavior analysis. Appl Sci 10(11):3978
30. Talukder S, Talukder Z (2020) A survey on malware detection and analysis tools. In: International Journal of Network Security and Its Applications (IJNSA), vol 12
31. Ugarte-Pedrero X, Graziano M, Balzarotti D (2019) A close look at a daily dataset of malware samples. ACM Trans Privacy Secur (TOPS) 22(1):1–30
32. Verdonck T, Baesens B, Óskarsdóttir M, et al (2021) Special issue on feature engineering editorial. In: Machine learning, pp 1–12
33. Vinayakumar R, et al (2019) Robust intelligent malware detection using deep learning. IEEE Access 7:46717–46738
34. Wu Q, Zhu X, Liu B (2021) A survey of android malware static detection technology based on machine learning. Mob Inf Syst 2021:1–18
35. Ye Y, et al (2017) A survey on malware detection using data mining techniques. ACM Comput Surv (CSUR) 50(3):1–40
36. Zhao Y, et al (2022) APIMatchmaker: matching the right APIs for supporting the development of Android apps. IEEE Trans Softw Eng 49(1):113–130

CHAPTER 4

Fast-Flux Service Networks: Architecture, Characteristics, and Detection Mechanisms

Basheer Al-Duwairi and Ahmed S. Shatnawi

4.1 Introduction

The Internet has witnessed an explosion in the kinds of tools available to attackers as well as attack techniques in recent years. Attackers continuously develop advanced tools and techniques to bypass defense technologies, conceal their identities, and evade detection. There are a lot of various tools that attackers can use to control systems they have compromised in target environments [1, 25, 40, 41]. These tools implement different ways to communicate across the network. This has resulted in a remarkable increase

B. Al-Duwairi (✉)
Department of Network Engineering and Security, Jordan University of Science and Technology, Irbid, Jordan
e-mail: basheer@just.edu.jo

A. S. Shatnawi
Department of Software Engineering, Jordan University of Science and Technology, Irbid, Jordan
e-mail: ahmedshatnawi@just.edu.jo

I. Almomani et al. (eds.), *Cyber Malware*, Security Informatics and Law Enforcement, https://doi.org/10.1007/978-3-031-34969-0_4

in the number of attacks with an increased level of sophistication at all protocol layers and exploiting vulnerabilities in protocols such as HTTPS, HTTP, and DNS.

The domain name system (DNS) is a critical Internet infrastructure component that primarily provides a mapping between domain names and IP addresses in addition to other services [26]. DNS is implemented as a distributed hierarchical database and is viewed as an entry point to the Internet as most Internet services depend mainly on resolving the IP addresses of domain names as a primary step in Internet access. For example, visiting a website or connecting to an FTP server is preceded by resolving that website's or FTP server's IP address. The central role of DNS in the operation of the Internet has attracted adversaries to take advantage of this core Internet infrastructure element to perform different types of malicious activities.

Most attacks involve the DNS in some way or another [21, 23]. However, certain attacks rely primarily on DNS. This includes performing DNS amplification attacks [6, 18, 31], DNS cache poisoning [20, 32], malware distribution [7, 22], and botnets [39–41]. In DNS amplification, an attacker instructs thousands of bot machines to send spoofed DNS queries to open DNS resolvers such that the target system is flooded with the corresponding DNS replies. In DNS cache poisoning, the attacker inserts a bogus DNS record in the DNS server cache such that Internet users are tricked into visiting a website controlled by the attacker. Fast-flux service networks (FFSNs) [14] have been used widely in recent years as a DNS-based mechanism to hide malware distribution servers or to provide robust communication with botnets' command and control (C&C) servers.

FFSNs provide shelter for web servers hosting malicious content. This technique has become one of the main techniques adopted by adversaries to provide highly available services while evading detection. This is especially important because cybersecurity professionals are equipped with tools and mechanisms that would allow them to identify malicious websites and blacklist them or shut them down when possible. Typically, a fast-flux network consists of thousands of bot machines known as flux agents that are configured to act as proxy nodes that relay traffic between end users and the mothership server hosting the malicious content [2, 14, 33]. Therefore, forming a protective layer prevents end users from directly reaching malicious servers.

FFSNs adopt techniques that are initially used in content distribution networks (CDNs) [8] and round-robin DNS (RRDNS) [9]. This is achieved by mapping the malicious server's domain name to multiple IP addresses selected from the pool of flux agents and changing over time, therefore increasing the chances that the origin server is reachable from some of the flux agents that are still running and have not been blacklisted yet. There has been growing concern in recent years about the increased adoption of fast-flux networks to protect phishing websites and botnets' C&C servers [19, 28, 44]. Clearly, identifying and shutting down these servers becomes more challenging when hosted in fast-flux service networks. This means that an effective and efficient approach to detecting the websites hosted in these networks is critical to addressing botnet-related attacks effectively.

Characterizing and understanding this type of malicious networks has received considerable research attention. Several research studies were conducted to fully understand the nature of this threat (e.g., [14, 17, 24, 37]), its new trends, and its role in hosting malicious phishing and spam campaigns. In addition, several fast-flux detection mechanisms (e.g., [2, 4, 12, 27, 33]) were proposed to efficiently detect fast-flux domain names based on characterizing features that can be obtained from different sources by active probing or passive probing. This chapter provides a detailed description of fast-flux networks focusing on their main characteristics and discussing major detection approaches. Also, it highlights new trends in fast-flux networks and identifies new features for fast-flux detection. However, it is important to mention that this chapter is not intended to provide a comprehensive survey of fast-flux detection approaches as in [46] and [5].

The rest of this chapter is organized as follows: Sect. 4.2 explains fast-flux networks and describes their main architecture. Section 4.3 discusses the main characteristics of fast-flux networks. Section 4.4 describes the main features that are typically used to detect fast-flux domains. Section 4.5 discusses fast-flux detection. Finally, Sect. 4.6 concludes the chapter.

4.2 Fast-Flux Service Networks

FFSNs were first introduced and described in detail by the Internet Honeyproject in 2007 [37]. Several research studies (e.g., [2, 14, 17, 33, 34]) were conducted to characterize and develop mechanisms for FFSNs detection. Most of these mechanisms rely on analyzing DNS traffic information that corresponds to fast-flux domains in order to characterize their behavior and identify their distinguishing features. In this regard, DNS records can be obtained actively by issuing DNS requests about domain names of fast-flux domains obtained from email spam campaigns and phishing archives or by analyzing passively collected DNS traffic traces.

FFSNs adopt techniques that are originally used in round-robin DNS and content distribution networks. Round-robin DNS and content distribution networks (CDNs) are two main techniques that web servers employ to achieve load balancing and high availability. In round-robin DNS [9], the authoritative domain name server of a certain domain name is configured to distribute the workload to multiple redundant web servers by mapping the hostname of the webserver to multiple IP addresses. This mapping keeps changing in a round-robin fashion. Each time a client issues a DNS query, the client may obtain a list of IP addresses for the given hostname in a different order. A content delivery network [8] is a service that accelerates Internet content delivery, therefore making websites much faster. This is achieved by reducing the distance between the user and the server providing the content by placing Content Delivery Network endpoints in as many locations worldwide as possible, reducing the amount of traffic that actually hits the server. In CDNs, the content is pushed to many geographically distributed servers. Global load balancing is achieved by providing the client with a set of IP addresses of nearby servers. For example, a user in the USA, who is trying to access a CDN-hosted website, sends a DNS query for that website and will get a reply with a set of IP addresses of servers that are hosted in nearby locations within the CDN.

FFSN [14, 29, 34] is a technique employed by botherders to hide their malicious webservers while providing high availability and resiliency. Figure 4.1 depicts the main steps of forming and operating these networks. Initially (Step 1), the botherder sets up the mothership server and configures it with a specific domain name (e.g., *myfastfluxdoamin.com*). This represents the server where the malicious content is hosted. It is also called the content server. Usually, a high-end machine with enough computing and storage resources is provisioned to serve as the mothership server. The mothership server usually hosts some sort of malicious content for the purpose of malware distribution, illegal pharmaceutical products sale, adult content, etc. Also, it can represent the command and control (C&C) server of another botnet. Then (step 2) the botherder registers the domain name *myfastfluxdomain.com* with a set of IP addresses that belong to a botnet controlled in advance by him/her. This step is crucial in the sense that the

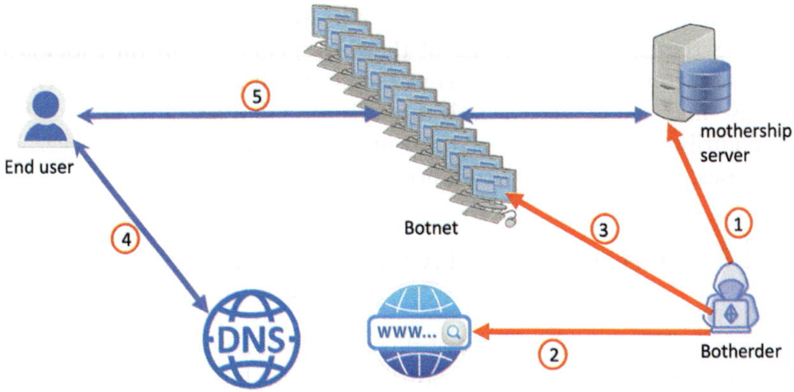

Fig. 4.1 Fast-flux service network life cycle

domain name of the mothership server does not map directly to the real IP address assigned to the mothership server.

The botherder configures the bot machines to act as proxies that forward traffic between end users and mothership servers (Step 3). Therefore, each bot machine that is part of this network is referred to as a *flux agent*. When visiting the website hosted by *myfastfluxdomain.com*, end users resolve the IP addresses of the mothership domain name *myfastfluxdomain.com* through the domain name system (Step 4). The domain name would be resolved to a set of IP addresses that belong to the botnet representing fast-flux agents. It is to be mentioned here that flux agents are mainly compromised machines with intermittent connectivity, limited computational power, and low to average bandwidth. Finally, the end user accesses the content through one of the flux agents returned by the DNS reply (Step 5).

It is clear that the botnet of flux agents forms a protective layer for the hidden malicious server. In order to increase the resilience of the network and to evade detection, the botherder keeps changing the domain name registration in a fast manner. This type of FFSNs is called a *single-flux*. There is a more sophisticated type of FFSNs called *double-flux* FFSNs, in which the botherder also changes the mapping between the authoritative name server of the FFSN and its IP addresses quickly, resulting in a constantly changing set of DNS servers, therefore providing a layer of protection for the FFSN' original authoritative name server. In this type of FFSNs, when a client resolves the IP addresses of a FFSN domain name, the DNS request is sent to one of the flux agents that in turn forward the request to the authoritative name server of that domain, and the DNS response message is relayed back through the same fast-flux agent. This may contribute to an additional delay in loading the website. However, in most cases, it goes without being noticed by Internet users.

4.3 Characteristics of Fast-Flux Service Networks

Classifying a domain name as a fast-flux domain or a non-fast-flux domain is a challenging problem because fast-flux domains look similar to domain names associated with CDN-hosted web servers in several aspects. In the following subsections, we illustrate the main differences between DNS *A* records associated with a fast-flux domain and DNS *A* records associated with a CDN-hosted domain name. Then, we identify main characteristics of fast-flux service networks based on previous research in this field.

4.3.1 Fast-Flux Domain Names Versus CDN-Hosted Domain Names

To illustrate the difference between a fast-flux domain name and a CDN-hosted domain name, let us consider the DNS name resolution of fast-flux domain *rgyui.top* and a CDN-hosted domain *timeline.com.* Figure 4.2 shows the DNS response message for domain *rgyui.top* and Fig. 4.3 shows

```
;; ANSWER SECTION:
rgyui.top.              75      IN      A       1.248.122.240
rgyui.top.              75      IN      A       211.171.233.129
rgyui.top.              75      IN      A       211.59.14.90
rgyui.top.              75      IN      A       138.36.3.134
rgyui.top.              75      IN      A       203.228.9.102
rgyui.top.              75      IN      A       190.140.74.43
rgyui.top.              75      IN      A       211.119.84.112
rgyui.top.              75      IN      A       189.165.26.224
rgyui.top.              75      IN      A       211.119.84.111
rgyui.top.              75      IN      A       210.92.250.133
```

Fig. 4.2 Output of the first dig of the fast-flux domain *rgyui.top* (performed on June 12 2022)

```
;; ANSWER SECTION:
timeline.com.          284     IN      A       52.4.38.70
timeline.com.          284     IN      A       52.1.173.203
timeline.com.          284     IN      A       52.1.147.205
timeline.com.          284     IN      A       52.5.181.79
timeline.com.          284     IN      A       52.4.175.111
timeline.com.          284     IN      A       52.4.145.119
timeline.com.          284     IN      A       52.6.46.142
timeline.com.          284     IN      A       52.1.119.170
timeline.com.          284     IN      A       52.4.225.124
timeline.com.          284     IN      A       52.6.3.192
timeline.com.          284     IN      A       52.0.16.118
timeline.com.          284     IN      A       52.4.240.221
```

Fig. 4.3 Output of dig of the CDN-hosted domain *timeline.com* (performed on June 12 2022)

Table 4.1 Country and ASN number of IP addresses of fast-flux domain *rgyui.top*

IP Address	*Country*	*ASN #*
1.248.122.240	Korea	AS9318
211.171.233.129	Korea	AS3786
211.59.14.90	Korea	AS9318
138.36.3.134	Brazil	AS264562
203.228.9.102	Korea	AS4766
190.140.74.43	Panama	AS18809
211.119.84.112	Korea	AS3786
189.165.26.224	Mexico	AS8151
211.119.84.111	Korea	AS3786
210.92.250.133	Korea	AS3786

the DNS response message for domain *timeline.com*. The common thing about these two domains is that they resolve to multiple IP addresses. However, a careful inspection of the response messages reveals major differences that would allow us to distinguish between fast-flux domains and CDN-hosted domains. Tables 4.1 and 4.2 show the country and ASN number of each IP address for both domains.

It is clear that IP addresses of a fast-flux domain are usually distributed in different countries and belong to several autonomous systems, while IP addresses of CDN-hosted domains are located in the same country (in most

Table 4.2 Country and ASN number of IP addresses of cdn-hosted domain *timeline.com*

IP Address	*Country*	*ASN #*
52.4.38.70	USA	AS14618
52.1.173.203	USA	AS14618
52.1.147.205	USA	AS14618
52.5.181.79	USA	AS14618
52.4.175.111	USA	AS14618
52.4.145.119	USA	AS14618
52.6.46.142	USA	AS14618
52.1.119.170	USA	AS14618
52.4.225.124	USA	AS14618
52.6.3.192	USA	AS14618
52.0.16.118	USA	AS14618
52.4.240.221	USA	AS14618

cases) and belong to the same autonomous system. This is expected because of the intrinsic behavior of fast-flux service networks where domain names are registered with botnet flux agents' IP addresses located in different countries and belonging to different autonomous systems. One of the main characteristics of fast-flux networks is the short TTL value assigned for their domain names compared to other domains. This is necessary to ensure the frequent and rapid change in mapping between IP addresses of flux agents and fast-flux domain names. As expected, performing another dig for the fast-flux domain name *rgyui.top* shortly after the first dig returned another set of IP addresses as shown in Fig. 4.4.

Fast-flux networks are characterized by frequent and fast mapping changes between domain names and IP addresses. As a result, a particular domain name would map to many IP addresses (selected from the pool of flux agents controlled by the botherder) over a short period of time. For example, the domain name *rgyui.top* maps to 17 distinct IP addresses based on two consecutive IP addresses. Of course, this number grows fast when performing more DNS queries for a longer period of time (Table 4.3).

In order to provide a reliable service and overcome the problem of blacklisting fast-flux domain names, fast-flux operators keep registering new domain names for their content servers. These domain names remain active for a short period and are assigned IP addresses from the pool of IP

```
;; ANSWER SECTION:
rgyui.top.          150     IN      A       187.170.250.215
rgyui.top.          150     IN      A       211.59.14.90
rgyui.top.          150     IN      A       190.140.74.43
rgyui.top.          150     IN      A       180.69.193.102
rgyui.top.          150     IN      A       196.200.111.5
rgyui.top.          150     IN      A       211.119.84.111
rgyui.top.          150     IN      A       109.98.58.98
rgyui.top.          150     IN      A       187.190.48.60
rgyui.top.          150     IN      A       148.0.95.36
rgyui.top.          150     IN      A       123.213.233.194
```

Fig. 4.4 Output of the first dig of the fast-flux domain *rgyui.top* (performed on June 12 2022)

Table 4.3 Country and ASN number of IP addresses of fast-flux domain *rgyui.top*

IP Address	*Country*	*ASN #*
187.170.250.215	Mexico	AS8151
211.59.14.90	Korea	AS9318
190.140.74.43	Panama	AS18809
180.69.193.102	Korea	AS9318
196.200.111.5	Eritrea	AS30987
211.119.84.111	Korea	AS3786
109.98.58.98	Romania	AS9050
187.190.48.60	Mexico	AS22884
148.0.95.36	Dominican Republic	AS6400
123.213.233.194	Korea	AS9318

addresses of fast-flux agents. Therefore, by monitoring DNS traffic traces over a long period, it is expected to observe that IP addresses belonging to specific fast-flux networks have been assigned to different fast-flux domains representing a malicious campaign. Figure 4.5 shows an example of such IP address reuse in fast-flux service networks, which is considered a standard practice. The figure shows a snapshot of the passive DNS replication history of IP address 183.78.205.92 obtained from VirusTotal [42]. The IP address was assigned to multiple fast-flux domain names on different dates.

183.78.205.92

Passive DNS Replication

Date resolved	Detections	Resolver	Domain
2022-06-15	6 / 92	VirusTotal	abababa.org
2022-06-14	6 / 92	VirusTotal	www.gerer.at
2022-06-12	5 / 92	VirusTotal	www.zfko.org
2022-06-12	9 / 92	VirusTotal	rgyui.top
2022-06-09	7 / 92	VirusTotal	esmic.at
2022-06-08	15 / 92	VirusTotal	xmhomestilesh.at
2022-06-03	8 / 92	VirusTotal	zfko.org

Fig. 4.5 A snapshot of passive DNS replication history of IP address 183.78.205.92 (obtained from VirusTotal on June 20th 2022)

4.3.2 Main Characteristics of Fast-Flux Service Networks

Several research studies have been conducted to identify malicious campaigns hosted by fast-flux service networks (e.g., [14, 24]). For example, the empirical study conducted in [24] showed that fast-flux service networks play a significant role in hosting scam campaigns. The study focused on monitoring DNS records of domain names hosting scam websites over a period of 1 month. The study revealed that fast-flux networks are usually shared among different spam campaigns. Figure 4.6 shows a visual representation of mapping between fast-flux domain names and their resolved addresses based on data used [2]. Each cluster represents a unique domain name and its associated IP addresses. This set of IP addresses and domain names is believed to belong to the same fast-flux service network.

Based on the discussion above and with reference to previous research studies that studied the problem of fast-flux networks (e.g., [14, 17, 24, 35], fast-flux networks are characterized by several characteristics that can be summarized as follows [2]:

- *Large number of IP addresses.* The A records included within a single DNS response message of a fast-flux domain are relatively large. Suppose one or more of the fast-flux agents that are associated with the IP addresses are down, a client. In that case, trying to access the mothership server of the associated domain name would automatically try another IP address (i.e., another agent) until it succeeds. Register-

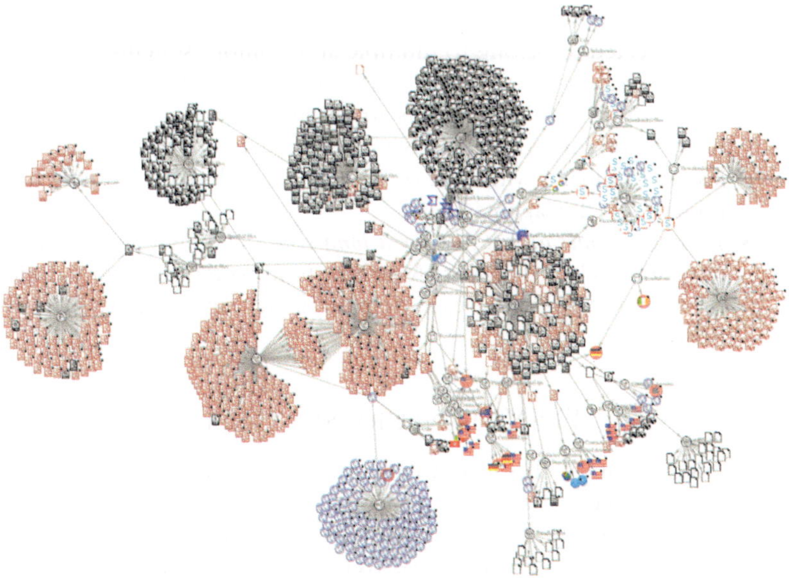

Fig. 4.6 Sample FF hostnames and their FFSNs observed in the study conducted in [2]

ing the domain name with many IP addresses provides high availability of the malicious server as it increases the probability that one of the flux agents is up and running.

- *Large IP growth.* To avoid blacklisting, mapping between a fast-flux domain and agent IP addresses keeps changing over time. Therefore, the number of IP addresses associated with a certain fast-flux domain becomes large.
- *Low TTL value.* Since the mapping between a domain name and IP addresses changes very fast in FFSNs, then the TTL values are kept low. This guarantees that the values expire soon after the fast-flux domain is resolved in order for users to obtain the new list of IP addresses.
- *Large number of autonomous systems.* The number of IP addresses that are returned in response to a DNS query for a fast-flux domain represents compromised machines that belong to different organizations

and Internet service providers. Therefore, these agents' IP addresses are expected to belong to multiple autonomous systems.

- *Large number of countries.* Previous studies showed that the IP addresses of fast-flux domains are usually located in a relatively large number of countries. This is expected since attackers register their fast-flux domains with a set of IP addresses selected randomly from a pool of fast-flux agents.
- *Domain names do not last for a long time.* The lifetime of a fast-flux domain is relatively short. Attackers tend to register many domains for their FFSNs, where each domain name remains active for a short period.

4.4 FFSNs Feature Set Collection

The detection of fast-flux service networks depends mainly on the information collected about the suspect domain names and their corresponding IP addresses. Information about domain names and their IP addresses can be collected from different sources passively and actively. The passive manner does not involve direct interaction with the source of information. Instead, it relies on analyzing passively collected information from traffic traces from strategic network locations. On the other hand, an active manner involves direct interaction with the source of information through request/reply-based protocols. Primary sources of information to be collected about suspect domain names include the domain name system, IP geolocation databases, and Internet-wide scanning projects. In the following subsections, we present these sources and highlight the main features obtained from each source.

4.4.1 Domain Name System-Based Features

The domain name system represents a primary source of information about domain names. When resolving a domain name into IP addresses, the DNS returns different types of resource records, such as A records and NS records. This includes the list of IP addresses for a domain name, the TTL value for the domain name, the number of IP addresses in each DNS reply, and the number of different IP addresses over a long period of time. This information can be obtained actively by probing DNS about suspect domains or passively by collecting DNS traffic traces using

Table 4.4 List of features extracted from DNS response message for fast-flux domain *rgyui.top* and legitimate domain *timeline.com*

Feature	*rgyui.top*	*timeline.com*
# IP addresses returned in one DNS lookup	10	12
domain name length	9	12
TTL value of DNS record	75	284
# of distinct ASNs for all IP addresses in a single DNS lookup	6	1
# unique IP addresses returned in all DNS lookups (IP address growth)	Additional information required	Additional information required

strategically placed network sensors while users are surfing the Internet. For example, with reference to the DNS response message of the fast-flux domain *rgyui.top* shown in Fig. 4.2 and legitimate domain name *timeline.com* shown in Fig. 4.3, the features shown in Table 4.4 can be extracted directly from their DNS response messages.

In active DNS probing, it is required to start with a list of suspect domain names that are usually obtained from email spam traps after extracting URLs embedded in spam emails and stripping domain names from them. A DNS lookup is then performed for each suspect domain name using the Unix *dig* utility or any other ns lookup tools. Main features are then extracted from DNS replies. A significant problem with this approach is that it results in a large number of DNS queries which may be suspected as a form of DDoS attack. In passive DNS probing, information about all domain names queried by users in an organizational network is collected passively. Collected DNS traffic traces are analyzed to filter suspect domain names based on specific criteria. While this approach does not incur additional DNS traffic, it deals with many DNS traffic traces that require significant computational and storage resources. However, it has the advantage of preventing false DNS replies that can be provided by attackers who might be controlling authoritative domain name servers while observing a large number of DNS queries. Moreover, it has the advantage of discovering fast-flux domains that could potentially appear in different malicious sources such as phishing emails, hacker forums, and online social networks.

4.4.2 *IP Geolocation-Based Features*

DNS-based features provide good insight into suspect domain names. However, additional information about suspect domain names and their IP addresses are usually required to accurately classify a domain name as a fast-flux domain name. This includes using (i) IP2location service to determine the location of IP addresses obtained through active or passive DNS probing. Having IP addresses scattered in different countries is an essential feature of fast-flux networks. Also, using (ii) IP to ASN lookup tool to determine the ASN number of each IP address as IP addresses that maps to specific fast-flux domain usually belong to different autonomous systems. For example, with reference to the DNS response message of the fast-flux domain *rgyui.top* shown in Fig. 4.2 and legitimate domain name *timeline.com* shown in Fig. 4.3, the features shown in Table 4.5 can be extracted by looking up their IP addresses in IP2location and IP2ASN databases.

Compared to legitimate domains, the IP addresses of FFSN domains exhibit a more uniform geographic distribution and a more widespread service relationship [43]. A framework to geolocalize fast-flux servers was proposed in [11]. The main objective of this framework was to determine the physical location of the fast-flux networks roots (mothership servers) based on network measurements. That was achieved through extensive network measurements from several vantage points distributed in the Internet. The framework was able to determine the physical location of fast-flux mothership servers within a distance of 100 km.

Table 4.5 List of features extracted from geolocation databases for fast-flux domain *rgyui.top* and legitimate domain *timeline.com*

Feature	*rgyui.top*	*timeline.com*
# of distinct ASNs for all IP addresses in a single DNS lookup	6	1
# of distinct ASNs for all IP addresses in a all DNS lookup	Additional information required	Additional information required
# of distinct countries	4	1
# of distinct countries for all IP addresses in a all DNS lookup	Additional information required	Additional information required

4.4.3 *Internet-Wide Scanning-Based Features*

Additional features of suspect domain IP addresses can be obtained from daily scans of IPv4 address space projects such as Censys [13] and Shodan [38]. Both Censys and Shodan are public search engines that provide information about devices connected to the Internet such as webcams, servers, routers, etc. The information provided by these search engines is collected by performing daily Internet wide scanning. For fast-flux detection, the certain information about suspect domain IP addresses can be collected. This includes the open ports on each scanned IP address, the service associated with each port number, and the operating system version. The premise here is that fast-flux agents corresponding to certain fast-flux domains do not necessarily have the same configuration. In contrast, it is expected that these flux agents would have heterogeneous network services running on them because each flux agent originally belonged to an end user who runs specific applications and services.

For illustration, we consider Shodan search results for two IP addresses (IP1, 211.171.233.126, and IP2, 222.232.238.243) selected arbitrarily from the set of IP addresses of fast-flux domain *rgyui.top* shown in Figs. 4.7 and 4.8, respectively. The search results show that port numbers 7, 17, 19, 8080, and 443 are open on the first IP address, while port numbers 80, 443, 4433, and 1434 are open on the second IP address. This difference in open port numbers is expected due to the fact that fast-flux agents belong to different end users. The count of unique port numbers that are open on all IP addresses associated with the fast-flux domain *rgyui.top* in a single

Fig. 4.7 Shodan search result for fast-flux domain *rgyui.top* IP address 211.171.233.126

Fig. 4.8 Shodan search result for fast-flux domain *rgyui.top* IP address 222.232.238.243

Fig. 4.9 Shodan search result for legitimate domain *timeline.com* IP address 52.2.173.203

DNS response message was eight. On the other hand, Shodan search results for two IP addresses (IP1, 52.1.173.203, and IP2, 52.6.3.192) selected arbitrarily from the set of IP addresses of fast the legitimate domain name *timeline.com* shown in Figs. 4.9 and 4.10, respectively. The search results show that both IP addresses have the same port numbers 80 and 443 open. In fact, all IP addresses that correspond to this domain have the same ports open. Table 4.6 shows the list of features extracted from Shodan.io for fast-flux domain *rgyui.top* and legitimate domain *timeline.com*.

4.4.4 *Active Delay Measurement-Based Features*

The fact that malicious servers hosted in fast-flux networks are accessed through flux agents rather than being accessed directly by end users implies that accessing a fast-flux domain incurs significantly more delay than access-

Fig. 4.10 Shodan search result for legitimate domain *timeline.com* IP address 52.1.119.170

Table 4.6 List of features extracted from Shodan.io for fast-flux domain *rgyui.top* and legitimate domain *timeline.com*

Feature	*rgyui.top*	*timeline.com*
# of distinct open ports	11	2
# IP addresses found in the database	8 out of 10	12 out of 12

ing a non-fast-flux domain. Flux agents work as proxy nodes that relay traffic between end users and mothership servers. Going through these agents takes additional processing and communication time. Typically, fast-flux agents are office, or home machines with limited computational power and intermittent Internet connectivity with low-speed Internet links [15]. In addition, it is expected that a flux agent's actual owner would run several applications and use the available bandwidth. This means that there is an excellent chance that connecting to a flux agent does not succeed from the first time or incurs additional overhead, resulting in additional delay in setting up the connection.

Performing active delay measurement indicates whether a domain name is a fast-flux domain name and may contribute to detecting fresh, fast-flux domains that did not appear yet on any blacklist or do not have enough DNS-related information to decide whether they are fast-flux domains or no. Here response time measurement variations can be observed spatially and temporally. Spatial variations are because fast-flux domain name maps to multiple IP addresses that are distributed in different locations and temporal variations to fluctuating workload on flux agents over time. In other words, performing delay measurement between an end user machine

Table 4.7 List of features commonly used by different fast-flux detection mechanisms

Feature	*Source(s)*	*Mode (Active/Passive)*
# IP addresses returned in one DNS lookup	DNS	Active/Passive
# unique IP addresses returned in all DNS lookups (IP address growth)	DNS	Active/Passive
# nameserver (NS) records in one single lookup	DNS	Active/Passive
TTL value of DNS record	DNS	Active/Passive
# of distinct ASNs for all IP addresses in a single DNS lookup	IP to ASN service	Active
# of distinct ASNs for all IP addresses in a all DNS lookup	IP to ASN service	Active
# of distinct countries	IP to location service	Active
# of distinct open ports	Internet wide scanning database	active
# of distinct open ports	Internet wide scanning database	active
# Response time difference	Active delay measurement	active

and a flux agent would be affected by the workload on that flux agent depending on the running applications and Internet usage.

Table 4.7 summarizes the main features commonly used by fast-flux detection mechanisms and shows the source of each feature and whether this feature can be obtained actively or passively. It is to be noted that various fast-flux detection mechanisms may use other features that are primarily derived from the list shown in this table.

4.5 Fast-Flux Detection

The main objective of fast-flux detection is to distinguish fast-flux domain names from non-fast-flux domain names. Typically, a domain name is considered a suspect domain name based on a combination of rules that are stemmed from fast-flux networks characteristics [45] such as (1) having short time to live value, (2) the domain name resolves to multiple IP addresses with scattered geographical distribution, and (3) frequent change of the set of resolved IP addresses returned in each query. Detection of fast-flux networks is generally achieved by analyzing certain information collected about suspect domain names from different sources as discussed in the previous section. Fast-flux detection mechanisms are generally characterized by the following attributes:

- *Detection mode:* This attribute refers to the mode of performing fast-flux detection which can be *offline* or *online* (i.e., in real time). Offline fast-flux detection mechanisms classify domain names into fast-flux domains and non-fast-flux domains by applying the detection algorithm on data sets collected in advance. On the other hand, online fast-flux detection mechanisms have the ability to classify domain names in real time.
- *Feature collection mode:* As explained in Sect. 4.4, fast-flux features can be collected either in active mode or passive mode. In active mode, specific features can be collected by issuing queries about the suspect domain name. While in passive mode, features about a suspect domain are obtained from traffic captured in a passive manner (e.g., while users are browsing the Internet).
- *Features used:* As explained in Sect. 4.4, there are four main types that are typically used by fast-flux detection mechanisms. This include DNS-based features, IP geolocation features, Internet wide scanning-based features, and active delay measurement-based features.
- *Classification algorithm:* This represents the core of any fast-flux detection mechanism. The majority of fast-flux detection mechanisms apply machine learning algorithms to classify domain names into fast-flux domain names and non-fast-flux domain names (e.g., [2, 4, 35, 36]). Generally, machine learning models are initially trained using labeled data sets containing a known fast-flux domains usually collected from blacklisted domains and legitimate domains usually collected from Alexa top domains list [3]. Feature selection algorithms are usually used to reduce the size of the feature set. Then the performance of the machine learning algorithm is evaluated in terms of accuracy, precision, false-positive rate, and false-negative rate. Machine learning-based FFSN detection mechanisms differ in several aspects that include FFSNs features used for classification, the mode of operation (i.e., whether the mechanism performs active or passive information collection), the machine learning algorithms used for classification, and the data set used for evaluation.

Table 4.8 Summary of main fast-flux detection mechanisms

Reference	*Detection mode*	*Feature collection mode*	*Features used*	*Classification algorithm*
Perdisci, R. et. al. [35],	offline	Passive	DNS-based	Machine learning algorithms
Hsu, F. H. et. al., [16]	online	Active	DNS-based and Active Delay measurement-based	statistical approach based on computing FastFlux score value
Al-Momani [4]	online	Active	DNS-based	Adaptive evolving fuzzy neural networks (EFuNN)
Perdisci, R. et. al. [35]	offline	Pasive	DNS-based, IP Geolocation-based	The C4.5 decision tree classifier
Hsu C-H et. al., [15]	online	Active	Delay Measurement –based	SVM classifier
Al-Duwairi B., et. al., [2]	online	passive	DNS based features, IP Geolocation-based, Internet Scanning –based	SVM (RBF kernel) classifier
Lombardo, P., et. al. [30],	offline	Active	DNS-based	Mathematical and data mining approach
Nagunwa, T. [33]	offline	Active + Passive	DNS-based, IP Geolocation-based, Delay- measurement based (A total of 83 features were introduced)	supervised ML techniques (e.g., SVM, DT, NB, LR)
Lin, H. T. et. al., [29]	online	Active	DNS-based, Delay-measurement based	Genetic algorithms
Zang, X. D., et. al [45]	online	Active + Pasive	DNS-based, IP Geolocation-based	different machine learning algorithms (e.g., SVM, C4.5, ELM)

Table 4.8 summarizes the main fast-flux detection mechanisms and compares them in terms of detection mode, feature collection mode, features used, and the core mechanism used.

In [35], the authors proposed a machine learning-based system, called FluxBuster, for FFSN detection. The high-level overview of FluxBuster is depicted in Fig. 4.11. Initially, DNS traffic traces are collected passively from different locations within an enterprise network. The traffic traces include mainly DNS A records that provide mapping between domain

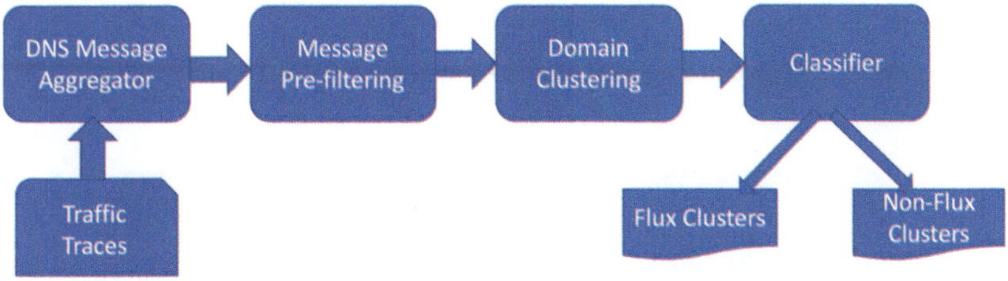

Fig. 4.11 High-level system overview of FluxBuster [35]

names and their resolved IP addresses. For a given domain name, the DNS Message Aggregator module aggregates information from all observed DNS messages corresponding to that domain during certain time interval. This includes the set of resolved IP addresses for that domain, the number of DNS queries observed during the monitoring time interval, and the average TTL values for collected A records. The aggregated DNS messages pass through the Message Pre-filtering module to filter out messages that correspond to unlikely fast-flux domains. Remaining domain names are processed by the Domain Clustering module, where domain names that share the same set of IP addresses are grouped together in one cluster. Finally, a machine learning-based classifier is used to classify each domain into fast-flux domain name or non-fast-flux domain name.

PASSVM [2] is a mechanism that performs online fast-flux detection of fast-flux domain names based on features extracted from the DNS response message itself, local Censys database, and local geolocalization database. The features include the number of IP addresses in the DNS response message, TTL value, domain name length, number of distinct countries where IP addresses are located, and number of distinct ASNs to which IP addresses belong. As depicted in Fig. 4.12, whenever a user visits a website, the A records of a suspect domain name received in a DNS reply message in response to a DNS query are analyzed, and the required features are obtained on the fly. Then, a decision is made on the fly whether the domain name is a fast-flux domain or a non-fast-flux domain by using the SVM machine learning algorithm.

Among the different features used in PASSVM, two new features extracted from the Censys database have significantly improved the accuracy of fast-flux detection. IP ratio: The ratio of the number of IP addresses returned from Censys to the number of IP addresses submitted in the

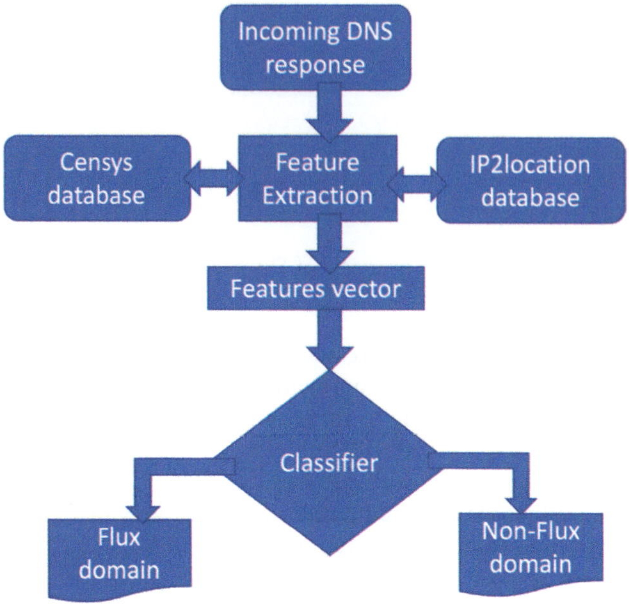

Fig. 4.12 High-level overview of PASSVM [2]

query. Ports: The number of distinct open port protocols for all IP addresses returned from the Censys search engine.

The authors in [16] proposed a fast-flux detection mechanism that is based on computing a FastFlux Score value. The system, called fast-flux domain detector (FFDD), consists of three major modules that include retriever, resolver, and recorder. The retriever performs active DNS probing using the UNIX dig utility for fast-flux domain names and legitimate domain names obtained from public sources. Also, it performs active delay measurements between the client machine and each of the resolved IP addresses after making the necessary formatting of the URL link. For each domain name, the resolver calculates the FF-Score value based on the response time measurements collected by the retriever and stored by the recorder module.

FFDD has a training phase and a testing phase. The main objective of the training phase is to determine the FF-Score of known fast flux and legitimate domain names. The FF-Score threshold value is selected manually in such a way as to distinguish fast-flux domains from other domains. In the detection phase, FFDD resolves the IP addresses for each suspect domain name by performing DNS lookups. Then, it performs active delay measurements to calculate the FF-Score for each domain. Here, the retriever sends 200 HTTP requests for each IP address. A suspect domain name is considered to be a fast-flux domain name if its FF-Score is larger than the threshold score determined in the training phase. A major problem with this approach is that it requires active interaction with suspect domains, where the client is supposed to issue large number of HTTP requests to each IP address associated with a suspect domain. This takes a lot of time and results in high traffic overhead.

The work presented in [14] was the first to present a detailed empirical study about FFSNs. It provided a comprehensive analysis of the threat of fast-flux networks and explained their main characteristics. Also, it developed a metric, called Flux-score, for fast-flux detection. This metric takes into account several parameters obtained from active probing of DNS servers. This includes the number of unique A records returned in all DNS lookups, the number of nameserver (NS) records in one single lookup, and the number of unique ASNs for all A records. Using linear programming [10], the optimal values of fast-flux score decision function parameters are determined.

4.6 Conclusion

Fast-flux service networks represent a major trend in the operation and management of botnets, malware distribution networks, and online spam/scam campaigns. In these campaigns, spammers flood email boxes of thousands of email users with advertisements about specific products or services (e.g., pharmaceutical, adult content, and phishing). The advertisements usually include hyperlinks to websites representing these campaigns' point-of-sale. Traditionally, spammers host the point of sale website using a domain name that maps to a single IP address or multiple IP addresses that remain constant for a considerable amount of time, which would allow defenders to quickly identify and blacklist these IP addresses, therefore denying access to spammers' websites. On the other hand, fast-flux service networks provide a layer of protection for point of sale website by mapping the website to multiple IP addresses that keep changing at a fast rate.

This chapter provided a detailed discussion of fast-flux service networks focusing on their architecture, operation, characteristics and detection mechanisms. Also, it highlighted their role in hosting online spam, scam, and phishing campaigns. There is a need to develop mechanisms to find the location of actual mothership servers, which would be necessary to shutdown malicious services hosted by them. Detection of zero-day fast-flux domains remains a challenging issue because of lack of information about these domains when they become active for the first time. It is possible that fast-flux service networks exhibit a behavior that deviate from the known fast-flux service networks characteristics. For example, fast-flux domain name can map to a single IP address that keeps changing quickly instead of mapping to multiple IP addresses. It is important to develop efficient mechanisms to detect such domain names.

References

1. Agarwal V, Mishra P, Kumar S, Pilli ES (2022) A review on attack and security tools at network layer of IoT. Optical and Wireless Technologies 2020:497–506
2. Al-Duwairi B, Jarrah M, Shatnawi AS (2021) PASSVM: a highly accurate fast flux detection system. Comput Secur 110:102431
3. Alexa—top sites. https://www.alexa.com/topsites (Accessed on 20/6/2022)
4. Almomani A (2018) Fast-flux hunter: a system for filtering online fast-flux botnet. Neural Comput Applic 29(7):483–493
5. Al-Nawasrah A, Almomani AA, Atawneh S, Alauthman M (2020) A survey of fast flux botnet detection with fast flux cloud computing. International Journal of Cloud Applications and Computing (IJCAC) 10(3):17–53
6. Anagnostopoulos M, Kambourakis G, Kopanos P, Louloudakis G, Gritzalis S (2013) DNS amplification attack revisited. Comput Secur 39:475–485
7. Aslan ÖA, Samet R (2020) A comprehensive review on malware detection approaches. IEEE Access 8:6249–6271
8. Bakiras S, Loukopoulos T (2005) Combining replica placement and caching techniques in content distribution networks. Comput Commun 28(9):1062–1073

9. Borkar GM, Pund MA, Jawade P (2011) Implementation of round robin policy in DNS for thresholding of distributed web server system. In: Proceedings of the international conference and workshop on emerging trends in technology, pp 198–201
10. Bradley PS, Mangasarian OL (2000) Massive data discrimination via linear support vector machines. Optim Methods Softw 13(1):1–10
11. Castelluccia C, Kaafar MA, Manils P, Perito D (2009) Geolocalization of proxied services and its application to fast-flux hidden servers. In: Proceedings of the 9th ACM SIGCOMM conference on Internet measurement, pp 184–189
12. Celik ZB, Oktug S (2013) Detection of Fast-Flux Networks using various DNS feature sets. In: 2013 IEEE symposium on computers and communications (ISCC). IEEE, New York, pp 000868–000873
13. Censys: attack surface management and data solutions. https://www.censys.io/ (Accessed on 20/6/2022)
14. Holz T, Gorecki C, Rieck K, Freiling FC (2008) Measuring and detecting fast-flux service networks. In: NDSS 2008
15. Hsu CH, Huang CY, Chen KT (2010) Fast-flux bot detection in real time. In: International workshop on recent advances in intrusion detection. Springer, Berlin, pp 464–483
16. Hsu FH, Wang CS, Hsu CH, Tso CK, Chen LH, Lin SH (2014) Detect fast-flux domains through response time differences. IEEE J Sel Areas Commun 32(10):1947–1956
17. Hu X, Knysz M, Shin KG (2011) Measurement and analysis of global IP-usage patterns of fast-flux botnets. In: 2011 Proceedings IEEE INFOCOM. IEEE, New York, pp 2633–2641
18. Ismail S, Hassen HR, Just M, Zantout H (2021) A review of amplification-based distributed denial of service attacks and their mitigation. Comput Secur 109:102380
19. Jiang H, Lin J (2021) Detect fast-flux domain name with DGA through IP fluctuation. International Journal of Network Security 23(1):88–96

20. Jin Y, Tomoishi M, Matsuura S (2019) A detection method against DNS cache poisoning attacks using machine learning techniques: work in progress. In: 2019 IEEE 18th international symposium on network computing and applications (NCA). IEEE, New York, pp 1–3
21. Khormali A, Park J, Alasmary H, Anwar A, Saad M, Mohaisen D (2021) Domain name system security and privacy: a contemporary survey. Comput Netw 185:107699
22. Kim S (2020) Anatomy on malware distribution networks. IEEE Access 8:73919–73930
23. Kim TH, Reeves D (2020) A survey of domain name system vulnerabilities and attacks. Journal of Surveillance, Security and Safety 1(1):34–60
24. Konte M, Feamster N, Jung J (2009) Dynamics of online scam hosting infrastructure. In: Moon SB, Teixeira R, Uhlig S (eds) Passive and active network measurement (PAM 2009)
25. Krishnamurthy P, Salehghaffari H, Duraisamy S, Karri R, Khorrami F (2019) Stealthy rootkits in smart grid controllers. In: 2019 IEEE 37th international conference on computer design (ICCD). IEEE, New York, pp 20–28
26. Kröhnke L, Jansen J, Vranken H (2018) Resilience of the Domain Name System: A case study of the. nl-domain. Comput Netw 139:136–150
27. Kumar SA, Xu B (2018) A machine learning based approach to detect malicious fast flux networks. In: 2018 IEEE symposium series on computational intelligence (SSCI). IEEE, New York, pp 1676–1683
28. Li W, Jin J, Lee JH (2019) Analysis of botnet domain names for IoT cybersecurity. IEEE Access 7:94658–94665
29. Lin HT, Lin YY, Chiang JW (2013) Genetic-based real-time fast-flux service networks detection. Comput Netw 57(2):501–513
30. Lombardo P, Saeli S, Bisio F, Bernardi D, Massa D (2018) Fast flux service network detection via data mining on passive DNS traffic. In: International conference on information security. Springer, Cham, pp 463–480

31. MacFarland DC, Shue CA, Kalafut AJ (2017) The best bang for the byte: characterizing the potential of DNS amplification attacks. Comput Netw 116:12–21
32. Man K, Qian Z, Wang Z, Zheng X, Huang Y, Duan H (2020) Dns cache poisoning attack reloaded: revolutions with side channels. In: Proceedings of the 2020 ACM SIGSAC conference on computer and communications security, pp 1337–1350
33. Nagunwa T, Kearney P, Fouad S (2022) A machine learning approach for detecting fast flux phishing hostnames. J Inf Secur Appl 65:103125
34. Passerini E, Paleari R, Martignoni L, Bruschi D (2008) Fluxor: detecting and monitoring fast-flux service networks. In: International conference on detection of intrusions and malware, and vulnerability assessment. Springer, Berlin, pp 186–206
35. Perdisci R, Corona I, Giacinto G (2012) Early detection of malicious flux networks via large-scale passive DNS traffic analysis. IEEE Trans Dependable Secure Comput 9(5):714–726
36. Rana S, Aksoy A (2021) Automated fast-flux detection using machine learning and genetic algorithms. In: IEEE INFOCOM 2021-IEEE conference on computer communications workshops (INFOCOM WKSHPS). IEEE, New York, pp 1–6
37. Salusky W, Danford R (2007) Know your enemy: fast-flux service networks. In: The Honeynet Project, pp 1–24
38. Shodan search engine. https://www.shodan.io/ (Accessed on 20/6/2022)
39. Silva SS, Silva RM, Pinto RC, Salles RM (2013) Botnets: a survey. Comput Netw 57(2):378–403
40. Thanh Vu SN, Stege M, El-Habr PI, Bang J, Dragoni N (2021) A survey on botnets: incentives, evolution, detection and current trends. Future Internet 13(8):198
41. Tuan TA, Long HV, Taniar D (2022) On detecting and classifying DGA botnets and their families. Comput Secur 113:102549
42. VirusTotal. https://www.virustotal.com/ (Accessed on 20/6/2022)

43. Wang HT, Mao CH, Wu KP, Lee HM (2012) Real-time fast-flux identification via localized spatial geolocation detection. In: 2012 IEEE 36th annual computer software and applications conference. IEEE, New York, pp 244–252
44. Williams J, King J, Smith B, Pouriyeh S, Shahriar H, Li L (2021) Phishing prevention using defense in depth. In: Advances in security, networks, and Internet of Things. Springer, Cham, pp 101–116
45. Zang XD, Gong J, Mo SH, Jakalan A, Ding DL (2018) Identifying fast-flux botnet with AGD names at the upper DNS hierarchy. IEEE Access 6:69713–69727
46. Zhou S (2015) A survey on fast-flux attacks. Inf Secur J: Global Perspect 24(4–6):79–97

CHAPTER 5

Efficient Graph-Based Malware Detection Using Minimized Kernel and SVM

Billy Tsouvalas and Dimitrios Serpanos

5.1 Introduction

Cyberattacks are continuously increasing, and the estimation of the total value at risk globally, due to these attacks, may reach 5.2 trillion USD until 2023 [37]. Cyberattacks employ different attack vectors, and a common goal is the insertion of malware to target systems. Malware is defined as a piece of software designed to cause damage or a program that performs an undesired action, whether it be to disrupt or gain unauthorized access to a system. Malware detection and classification is a hard task that becomes increasingly difficult when we consider the high production rate of new malicious executables and re-purposing of previously deployed

B. Tsouvalas (✉)
Stony Brook University, Stony Brook, NY, USA
e-mail: vtsouvalas@cs.stonybrook.edu

D. Serpanos
Computer Technology Institute and Press DIOPHANTUS, University of Patras, Patras, Greece
e-mail: serpanos@cti.gr

I. Almomani et al. (eds.), *Cyber Malware*, Security Informatics and Law Enforcement, https://doi.org/10.1007/978-3-031-34969-0_5

ones (malware variants). This leads to a malware landscape that consists of diverse attack strategies and a wide range of vulnerability targets. Malware is divided into categories, such as viruses, Trojans, worms, rootkits, ransomware, adware, etc., based on the method employed to carry out the attack, to distribute copies (if any) and, in general, the techniques used for a security breach. Considering that malware-related cybercrime and breaches account for 28% of all cyberattacks [64], it is apparent that malware attacks are a problem of paramount importance and consequence, and furthermore, malware analysis and detection mechanisms are essential against cybercrime.

Malware analysis is performed with two different types of analysis: static and dynamic [27]. Static analysis examines a software sample without executing it, while in dynamic analysis the sample is executed in a secure environment, typically a virtual machine named sandbox, in order to collect runtime execution data. Typical static analysis methods include signatures, where a predefined database of unique digital malware signatures enables threat recognition through signature matching, disassembly methods, and code analysis, where a sample is reverse-engineered, using a disassembler or a decompiler, in order to examine the code and classify executable. Dynamic analysis is behavioral and allows for additional information extraction, because the runtime behavior of the sample is explored. Dynamic analysis is more advantageous, because it enables collection of significant information, observes actual sample execution, and can evaluate more aspects of the sample. However, dynamic analysis is significantly more costly and power demanding, because it requires the setup of a virtual machine and its teardown, in addition to the execution of the sample. Considering the high computational cost of dynamic analysis, significant effort is spent to develop highly effective static analysis solutions to the malware problem.

We introduce an efficient static analysis method, which includes disassembling an executable sample, extraction of the API call graph, and evaluating the sample based on graph analysis. Our analysis exploits machine learning methods, considering that such methods are widely employed in malware detection, especially in classification [25, 40, 47, 55, 67, 72, 74], achieving very high accuracy providing promising results. In our work, we use support vector machine (SVM) for the binary classification part of the process, which is a classification algorithm that accounts for 29% of all learning schemes applied to malware detection [56]. For the analysis, we use models of API call graphs of both benign and malicious executable samples. In order to reduce the size of the statically constructed API call graphs, we perform a graph abstraction, which enables us to vectorize the comparison among samples. Specifically, we vectorize the graph comparison problem using a minimized version of a random walk graph kernel, which is, in turn, used as the input of the SVM. We evaluate the results of this approach for an unweighted and a weighted version of the API call graph, and we employ a dataset that contains samples that we have collected. The curated dataset contains malware, collected from acknowledged malware sources, such as VirusShare [2], and benign software, composed of popular benign software such as operating system installation files and version control managers, i.e., Windows [3] or Git [4]. Considering the dependence of malware on the operating system that is required for sample execution, we have focused on Windows samples, i.e., the executable is a Windows executable and the API graph includes its calls to the Windows API; the method can be easily implemented for alternative operating systems (Unix, Linux, MacOS, etc.).

Regarding the dataset, we note that, although malware detection and classification are an active field of research and development, there do not exist benchmark datasets containing executable samples. The benchmark datasets employed in relevant literature and used by the community are mainly curated by the industry and, typically, contain only extracted features of malicious executables, instead of the executable samples themselves [14, 24, 29, 49]. Independently of the lack of benchmarks, there is not even a single dataset of executable samples that is used widely—as a reference point—to compare different malware detection methods and frameworks. Importantly, many malware detection tools employ proprietary and unavailable datasets for experiments and evaluation. This leads to results that are not reproducable and thus not comparable. Furthermore, many datasets used in malware detection experiments contain very few samples, limiting the ability for safe conclusions about their effectiveness and performance [17]. For this reason, we collect and curate our own dataset that contains benign and malicious executable samples. Importantly, we analyze the collected data and provide an evaluation on the similarity of the samples, demonstrating the dataset's diversity.

The paper's contribution is a method for efficient graph-comparison of API call graphs exploiting the problem's constraints, in order to achieve an efficient, high-performance malware detection method. More explicitly, our contributions are:

- introducing a method to construct abstract API call graphs taking into account problem constraints and enabling calculation of random walk graph kernels more efficiently than current methods (our method achieves $O(n^3)$ complexity relatively to $O(n^6)$, where n is the number of nodes in the graph);
- achieving the same accuracy levels as similar efforts [21] by using a substantially smaller dataset;
- achieving higher accuracy levels by introducing weighted API call graphs.

The chapter is organized as follows: subsequent to the presented Introduction of Sects. 5.1 and 5.2 discusses the related work and state-of-the-art of the field, Sect. 5.3 contains extensive description of the graph-based modeling and malware classification scheme, Sect. 5.4 presents the experimental settings and results of the malware classification mechanism, and Sect. 5.5 concludes the chapter.

5.2 Related Work

There are several approaches that address malware detection utilizing the calls that a software sample makes to the operating system, especially for the Windows API. API calls of an executable provide information about the resources it needs from the operating system; the API call sequence is a fundamental behavioral characteristic of an executable, because it reflects its control flow and possible execution paths. In this direction, efforts have targeted to identify API call sequence differences between benign and malicious executables and to exploit these differences to formulate detection schemes. API call sequences are a widely used feature for several detection methods; such methods consider API call selection, frequency, and ordering/sequential characteristics [41], identify distinct API call sequences and use n-gram models for the sequence length selection [9, 63, 72], use feature extraction algorithms based on the behavioral analysis of the various API call sequences [61, 62], and develop metrics to measure the similarity between malware through alignment techniques [19] or word embedding and clustering schemes [13].

Other approaches employing API call sequences focus on searching for the longest common subsequence of API calls extracted using dynamic analysis [38], while language-based models have also been adapted to API call sequences to measure the similarity between executables and thus conclude on their nature [71]. API calls provide crucial information to analyze a sample's runtime behavior and, thus, have been used for analysis in alternative approaches that do not take into account API call sequences. These approaches include classification of API functions based on their behavior and the potential malicious intent of the software [10]. Features such as the API call names and input arguments have proven effective in detection schemes [50, 51]. Furthermore, API call-related features have been employed for obfuscated malware detection [9], while the frequency of API call usage has also been employed for classification purposes [11]. Other state-of-the-art approaches have employed the appearance of API calls in malicious and benign software as a means of categorization [52]. In mobile systems security, where there is high availability of open Android application packages (APK) and mobile malware samples, several efforts for mobile malware detection that employ API calls and Android permission requests have been quite successful [12, 22, 45].

Graph-based malware analysis is widespread, considering the typical modeling of flows in programs. Graph-based detection schemes that employ static analysis have proven effective and achieve high performance, either by focusing on function call graphs and clustering mechanisms [30] or by following more general approaches such as API calls and dynamic link libraries (DLLs) relations modeled into a heterogeneous information networks (HIN) [26]. Moreover, in mobile malware detection, employing different characteristics of statically extracted API calls, such as names, frequency of appearance, or other sequence characteristics, has yielded encouraging results [41].

Dynamic analysis often employs graph-based schemes, which include API or system calls. This is carried out in the context of dynamically extracted API call sequences modeled into API call graphs [18] or by utilizing dynamic taint analysis to produce system call and dependency graphs and producing similarity metrics between samples in order to detect and classify malware [44].

A promising approach to malware detection based on API call graphs that involves static extraction of API calls, modeling of each executable with an API call graph, and comparison of these graphs using a kernel has provided very good results [21]. Our approach is analogous, and, in contrast to the conventional one, it exploits the constraints of the problem, and, more specifically, it addresses the possible size reduction during the subgraph comparison of the labeled API call graphs. This leads to even higher accuracy, faster, and with a smaller dataset.

In the context of API and system call-based malware classification, machine learning (ML) and deep learning (DL) techniques have been widely adopted. ML approaches include clustering algorithms such as k-nearest neighbors [68] and decision-based methods [34], while DL frameworks have shown promising results, with approaches employing autoencoders [32, 35] and convolutional neural networks [42, 46, 48]. Furthermore, combinations of ML and DL components, using graph convolutional networks [16] and stacked autoencoders [70] along with a variety of classifiers, have demonstrated elevated detection performance. Although there is a wide range of classification methods employed in API-based malware research, SVM techniques still hold a prevalent position in the field [22, 22, 45, 69].

Support vector machine [20] is a machine learning supervised algorithm used widely for malware binary classification (malware/benign). Its extensive use originates from the fact that it allows fine-tuning of its parameters in order to avoid overfitting and underfitting; this makes SVM advantageous over alternative machine learning algorithms for malware detection. Representative frameworks that employ SVM include (i) n-gram schemes on API call sequences [9, 59] where classification takes as input vectors of sample extracted features, (ii) text-mining approaches on API call sequences [58], and (iii) API call flow graph vectorization and feature extraction schemes [18] where features are extracted using data mining techniques and used for training.

We employ SVM for the classification part of our method after having transformed the API call graph comparison into a feature vector through use of an appropriate minimized random walk graph kernel; the kernel serves as a similarity metric for two graphs.

The lack of available benchmark malware datasets, which contain both malicious and benign executables, is a significant limitation in malware analysis research and development. Available and recent malware datasets typically contain extracted features of malicious and benign executables, but not the executable samples themselves. The main feature-based benchmark datasets are summarized in Table 5.1. Although datasets with extracted features enable analyses with machine learning frameworks, they are inappropriate for methods that analyze software samples prior to classification. The unavailability of public and free datasets with samples is also due to intellectual property constraints [29]. Specifically, the constraints are for benign samples which may be parts of proprietary software. Malicious executable samples can be collected from dedicated malware repositories,

Table 5.1 Feature-based benchmark malware datasets

Dataset	*Size*	*Feature extraction*	*Executable samples availability*
Ember [14]	1.1M	LIEF [60]	None
SoReL-20M [29]	20M	Ember [14] features and PE metadata	9, 919, 251 malware binary samples—No benign
BIG 2015 [49]	> 20*k*	Disassembly and Bytecode	Hexdump conversion necessary—No benign

such as VirusTotal [5] and VirusShare [2], using their hash. Due to these limitations, most efforts that require the availability of executable samples employ proprietary or privately collected datasets, and, thus, their results are not reproducible, and there is no ability for independent comparisons. Since our approach requires analysis of executable samples prior to classification, we have created our own dataset, collecting benign and malicious executable samples; to demonstrate the dissimilarity and diversity of the collected samples, we provide a similarity metric for both sets of samples as we describe in Sect. 5.4.

5.3 API Call Graph-Based Analysis Framework

We introduce a graph-based detection scheme based on static analysis of executables and a machine learning technique, specifically support vector machines (SVMs). Our scheme follows trends of recently proposed schemes [21, 23, 43, 73] and differs from alternatives in the construction of the final graph used for the classification which is based on SVM.

Our method is composed of four stages (steps), considering the existence of a labeled dataset DS of benign and malicious executables. When a new sample (executable) S, not included in DS, is processed for classification, the four processing stages are the following:

1. extraction of the API call graph of S;
2. extraction of an abstract API call graph of S;
3. for every executable D in DS, calculation and reduction of a graph kernel
4. classification

The extraction of the API call graph (ACG) for an executable S is achieved through disassembly of the executable, extraction of its API calls to the operating system, extraction of API call sequencing information, and subsequent construction of the ACG based on the collected data. Then, in the second step, the ACG is processed, reducing the ACG and calculating an abstract API call graph (AACG) of *S*, which has a smaller size. In the third step, the AACG is used to make pairwise comparisons of the AACG of *S* with the AACGs of all the executables in *DS*. These pairwise comparisons are calculations of an appropriate graph kernel [39], which effectively vectorizes the AACGs and renders the comparison data appropriate for use in classification. Finally, in the fourth step, SVM-based classification inputs the vectorized comparisons to classify the sample *S* as malicious or benign.

In the following subsections, we detail our method based on the implementation of our method, which focuses on Windows executables and binary classification (malicious or benign).

5.3.1 Extraction of API Call Graph

We disassemble *S* using the open source program Ghidra [1]. Ghidra provides a wide range of operations, such as disassembly and decompilation while enabling scripting and graph representation of data related to resources identified in the disassembled code. An important capability of Ghidra is that its API can be used to develop custom code to perform desired procedures, such as the resource extraction that is required in our method. In our method implementation, we extract the API calls that *S* makes and their connectivity information in order to construct API call graphs.

A successful Ghidra analysis of *S* establishes that *S* is correctly disassembled and its code can be represented in an assembly language. From the assembly code, we extract basic information of the executable, such as the used registers, the called functions, the included API calls, and other relevant data. Based on the assembly code, Ghidra calculates codeblocks, which are bundles of disassembled code that Ghidra relates distinct actions to. Effectively, codeblocks are considered as internal functions of *S*.

5.3.2 Extraction of Abstract API Call Graph

Using the extracted codeblocks, we construct the control flow graph (CFG) of *S* denoted as $G = (N, E)$, where N is the set of nodes and E is the set of edges. *G* is a directed graph, where nodes correspond to

Table 5.2 AACG size reduction

Sample	*Label*	*ACG nodes*	*AACG nodes*	*Gain*
VSd0595f53a68e289e55c9aa37546c6c89	Malware	956	5	191.2
VS27c17e3b1111fc5c3d4f6d779b15d4da	Malware	911	5	182.2
VScc755bfe842d44f14d87b77848e4ed6d	Malware	538	4	134.5
GitHub.Authentication.exe	Benign	131	4	32.75
FlashPlayerApp.exe	Benign	2688	206	13.05

S's codeblocks and edges connect codeblocks in sequence to model the possible execution paths. Thus, two nodes n_i and n_j represent two distinct codeblocks, and an edge (n_i, n_j) demonstrates that the execution of the codeblock of n_i can be followed by the execution of the codeblock of n_j in an execution path of S.

The API call graph (ACG) of sample S is the directed graph that is extracted from the CFG of S, by replacing each node with the API calls that the specific Codeblock leads to. For example, if the codeblock represented with node n_i leads to API call f_1 and the codeblock of node n_j leads to API call f_2, the ACG includes nodes (n_i, f_1) and (n_j, f_2), respectively. The ACG is denoted as $G_{APICall} = (N_{API}, E_{API})$. We reiterate that the nodes of a CFG are codeblocks, which have been extracted by the disassembly process. These codeblocks may lead to zero, one, or more API calls. Given the fact that a single API call may be reached from several points during the execution of the software, the size of the ACG may exceed the size of the CFG in terms of nodes and edges. Furthermore, considering that a codeblock of S may lead to several API calls, we expect the size of the ACG to be greater than that of the CFG.

Our method targets to exploit pairwise comparisons of sample ACGs for classification with SVM. The speed of pairwise calculations is critical and depends on the size of the compared graphs. For this, considering that ACGs have large sizes, we calculate a reduced size abstract API call graph (AACG) to enable faster comparisons.

An AACG is an undirected graph that is constructed from an ACG by merging all ACG nodes that correspond to the same API call. The AACG nodes are connected with undirected links merging the corresponding links in the ACG. Thus, the AACG's node set is equal to the set of distinct API calls that are used by the executable sample. In Table 5.2, we provide a few examples of graph size reduction achieved by the AACG manipulation.

In addition to the unweighted AACG described, we also consider the case of a weighted AACG. The weights are the transition frequencies between the connected API calls.

5.3.3 *Calculation and Reduction of a Graph Kernel*

Our method is based on pairwise comparisons of executable samples, with the objective to calculate a measure of their similarity. We calculate similarity exploiting a random walk graph kernel, which vectorizes graph comparison and enables classification exploiting vector-based techniques, such as SVM. Our choice for the kernel originates from the need to use the similarity measure as input to the SVM for classification.

A graph kernel is a kernel function, i.e., a generalized dot product function [54], that calculates the inner product of two graphs and provides a measure of similarity between them [8, 15]. Thus, a graph kernel is suitable for kernelized learning and classification algorithms such as SVMs. In our method, we use a special case of graph kernels, a random walk graph kernel.

Random walk graph kernels (RWGK) for two graphs G_1 and G_2 calculate random walks on the two graphs and count the number of matching walks [66]. Importantly, it is proven that performing random walks simultaneously on a pair of graphs G_1 and G_2 is equivalent to performing a random walk on the direct product graph $G_1 \times G_2$ of the two graphs [33].

Given two graphs $G_1 = (N_1, E_1)$ and $G_2 = (N_2, E_2)$, their direct product graph $G_\times = (N_\times, E_\times)$ is a graph over all possible pairs of nodes from G_1 and G_2, where two nodes of $G_\times$ are neighboring if and only if the corresponding nodes in G_1 and G_2 are neighbors in both graphs.

In the following, we consider the representation of graphs with their adjacency matrices: A_1 and A_2 are the adjacency matrices of G_1 and G_2, respectively. The adjacency matrix of $G_\times$ is

$$A_\times = A_1 \times A_2 \tag{5.1}$$

In order to calculate a random walk in a graph, we need to define distribution probabilities for the starting and stopping point of the random walk; we denote these probabilities p and q, respectively [28, 66]. Considering the two graphs, G_1 and G_2, their starting and stopping distributions are

p_1, q_1, and p_2, q_2, respectively. Considering these distributions for G_1 and G_2, the equivalent probability distributions of the direct product graph are

$$p_\times = p_1 \otimes p_2 \quad and \quad q_\times = q_1 \otimes q_2 \tag{5.2}$$

The Random walk graph kernel for the pair G_1 and G_2 is defined as

$$\kappa(G_1, G_2) = \sum_{k=0}^{T} \mu(k) q_\times^\intercal A_\times^k p_\times \tag{5.3}$$

where:

- $A_\times$ is the adjacency matrix of $G_\times$, and, thus, A^k represents the probability of simultaneous k-length random walks on G_1 and G_2;
- $p_\times$ and $q_\times$ are the initial and stopping probability distributions of $G_\times$, respectively (as described in (2));
- T is the maximum length of a random walk;
- $\mu(k) = \lambda^k \in [0, 1]$ is a coefficient that controls the importance of length in random walks which we use to ensure that the sum converges and the kernel value is well defined [66].

In the implementation of our method, we use $p_1 = q_1 = 1/|N_1|$ and $p_2 = q_2 = 1/|N_2|$, i.e., uniform distributions over the nodes of G_1 and G_2, as commonly used [21, 28]. For the extracted abstract API call graphs, we note that:

- $A_\times$ in the kernel definition refers to an unweighted graph, while in the case of weighted graph, $A_\times$ is replaced with matrix $W_\times$ which contains the edge weights;
- the nodes of the graphs in our method are labeled, where the node labels are the names of the corresponding distinct API calls. Thus, $W_\times = A_\times$, leading to Eq. (5.3) [28, 57, 66].

Based on the above, the kernel definition becomes

$$\kappa(G_1, G_2) = q_\times^\intercal (\mathbf{I} - \lambda A_\times)^{-1} p_\times \tag{5.4}$$

In Eq. (5.4), we employ the generalized definition of the random walk graph kernel for labeled graphs [28, 57, 66].

The calculation of the kernel is equivalent to inverting $(\mathbf{I} - \lambda A_{\times})$ as derived from Eq. (5.4). Considering that the complexity to invert a matrix is $\mathcal{O}(n^3)$, where n is the matrix dimension (rows/columns in the adjacency matrix), the computation to invert $(\mathbf{I} - \lambda A_{\times})$ is $\mathcal{O}(n^6)$ [36, 66]. In our method, we calculate the kernel with complexity of $\mathcal{O}(n^3)$ by pre-processing the adjacency matrices and performing the kernel calculation employing decomposition into Kronecker products and the Sylvester method [65]. To achieve this, we exploit properties of the AACGs, eliminating unnecessary nodes and reducing the adjacency matrices involved in the kernel computation to equal sizes. Specifically, considering that AACGs have uniquely labeled nodes (the labels are the API calls) and that only common nodes contribute to the results, we can eliminate all the nodes that are different between the two compared graphs. Thus, the resulting matrices have the same dimension and the same labels. The employment of equal size matrices in the kernel computation enables decomposition into Kronecker products and adoption of the Sylvester method [65], which leads to the lower complexity of $\mathcal{O}(n^3)$. Importantly, the elimination of the unnecessary nodes from the direct product graph and the corresponding costly computations reduce significantly the running time of the graph comparison method [36].

The random walk graph kernel is a measure of similarity between two graphs [8, 15]. We compute kernel values and evaluate the similarity of all possible pairs of graphs. Thus, for every executable sample, we compute a kernel value for every other executable sample in DS. For a dataset of M executable samples, we have a total of $(M - 1)^2$ kernel values, overall. Given these kernel values, which correspond to the all pairwise comparisons of samples in a dataset, we can directly use the kernel, as input to the SVM classifier. The principle of kernelization of the data is known as the kernel trick [31, 53].

After their calculation, kernel values are normalized as follows:

$$\hat{\kappa}(G_1, G_2) = \frac{\kappa(G_1, G_2)}{max(\kappa(G_1, G_1), \kappa(G_2, G_2))} \tag{5.5}$$

Normalization leads to kernel values in the range $\{0, 1\}$, where 1 is the result when a graph is compared to itself.

5.3.4 *Classification*

Classification is performed using all kernel values of the executable pairs (S, D_i), where $D_i \in DS$, as well as all kernel values of the executable pairs (D_k, D_l), where $D_k, D_l \in DS$; these values are M^2, where $M = |DS|$. These kernel values are used for a support vector machine (SVM) classification scheme.

The training of the SVM is performed using the set of the M vectors, where each vector has the form $[(x_{11}, y_1), \dots, (x_{1n}, y_1)]$, $[(x_{21}, y_2), \dots, (x_{2n}, y_2)]$, ..., $[(x_{n1}, y_n), \dots, (x_{nn}, y_n)]$, where x_{ij} represents the kernel value of the comparison of samples S_i and S_j, and y_i is the class label, which is an integer value a with $a = -1$ or $a = 1$ for malicious and benign executables, respectively.

5.4 Experiments and Testing

We evaluate our method in two directions: (a) the effectiveness of the kernel results and (b) the SVM classification.

In the first direction, we conduct experiments for both weighted and unweighted AACGs, calculating and comparing kernel values for sample pairs. We analyze separately benign and malware samples, to establish metrics for each category, and then we compare the categories to evaluate their effective separation using kernel values as a metric. Specifically, we conduct the following measurements:

1. Benign-Benign: Since the kernel is a measure of similarity, we demonstrate the extent to which the benign samples are similar among them. In this context, a small kernel value demonstrates the dissimilarity of the dataset and the reliability of the benign sample collection.
2. Malware-Malware: In the same manner as for the benign samples, we observe the kernel values (similarity measurements) of the malicious executables and conclude on the reliability of the malicious sample collection.
3. Benign-Malware: The kernel value of the comparison of benign and malicious samples is a preliminary classification metric, which demonstrates how similar or different the benign and malicious samples are. In this context, we can observe the inner workings of a classifier.

Using the computed kernel values for the benign and the malicious samples, we also evaluate binary classification using SVM. Considering as baseline the case of the unweighted AACG, we evaluate the weighted AACG approach.

5.4.1 *Dataset*

We have created our own dataset for the experimentation and testing of our method, due to lack of appropriate public datasets. Focusing on the Windows operating system, the dataset includes both benign and malicious Windows executable files, containing only labeled executable samples. The included benign samples are installation and support files and have been collected from trusted sources such as Windows [3], Git [4], Cygwin [6], and Codeblocks [7]. The dataset includes 567 benign executables, with size ranging from several hundred KB to several MB. The malware samples have been drawn from VirusShare [2], which is a website that provides malware samples for academic and scientific purposes. We collected 827 malicious executables, with sizes ranging from several hundred KB to several MB.

5.4.2 *Evaluation of Kernel Effectiveness*

We evaluate the effectiveness of the kernel use as a metric and the diversity of the collected executable samples by analyzing the dissimilarity of benign and malicious samples independently, as well as among these two sample categories. Analyzing similarity of all benign samples and all malware samples independently shows us that the samples of these two subsets of the dataset are not similar; thus, the results of the method are reliable.

5.4.2.1 Unweighted API Call Graph

For the 567 benign dataset samples, we calculate the kernel values for subsets of 50, 100, 150, 200, 250, and 300 benign samples; we perform 3 calculations per subset size, drawing samples with uniform probability among the 567 benign samples for each calculation. Analogous calculations are made for the 827 malware samples. The average kernel value for each dataset size appears in Table 5.3. Furthermore, Fig. 5.1 plots all results including a histogram of the kernel value distribution at the right of each plot. As the results show, the maximum average kernel value, i.e.,

Table 5.3 Benign-benign and malware-malware normalized kernel values average of unweighted API call graph for the different dataset sizes

Subset size	*50*	*100*	*150*	*200*	*250*	*300*
Benign—Benign	28.34	29.09	**32**	29.35	28.86	28.68
Malware—Malware	22.86	22.92	**28.58**	20.53	25.39	22.21

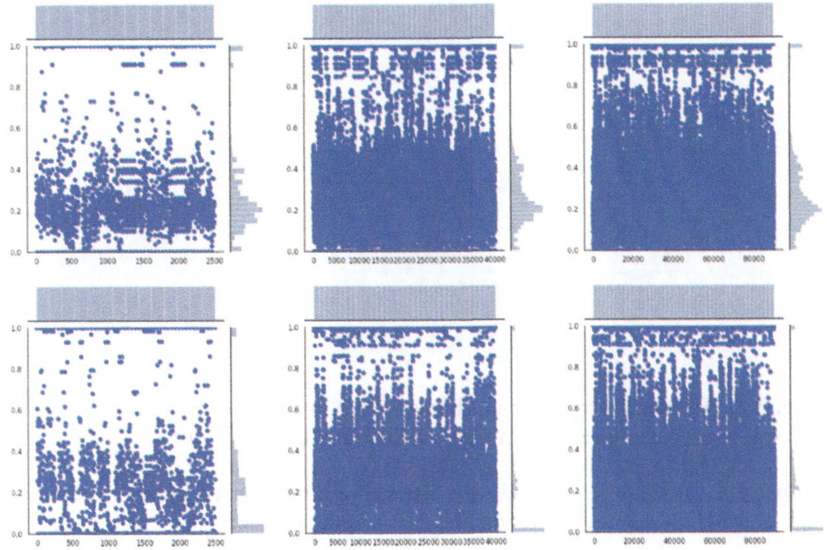

Fig. 5.1 Unweighted graph: benign-benign (top), malware-malware (bottom) kernel values for dataset size: (from left to right) 50, 200, and 300

the maximum average similarity metric, is 32% for benign and 28.58% for malware samples, respectively. This indicates that the similarity metric between benign and malicious samples is sufficiently different.

5.4.2.2 Weighted API Call Graph

We perform the same procedure for the weighted graph approach, and we demonstrate that the maximum average kernel value for benign and malware samples does not surpass 32%. We present these results in Table 5.4 and Fig. 5.2.

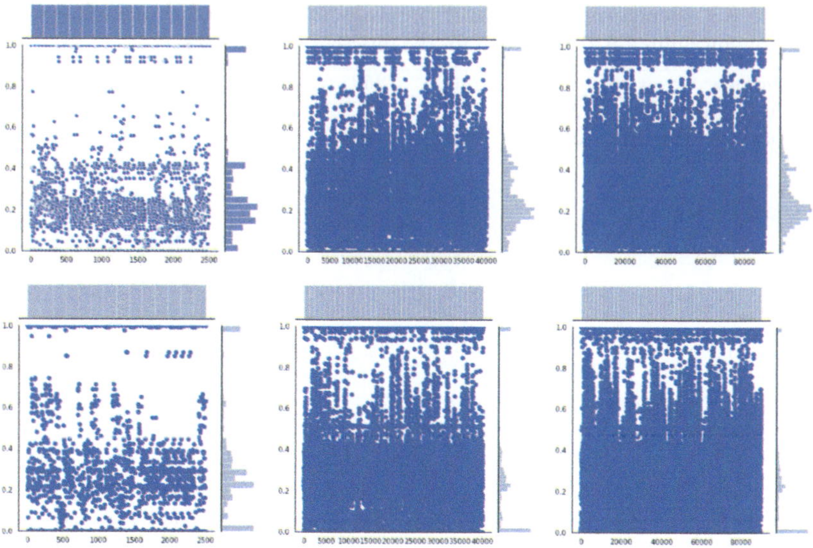

Fig. 5.2 Weighted graph: benign-benign (top), malware-malware (bottom) kernel values for dataset size: (from left to right) 50, 200, and 300

Table 5.4 Benign-benign and malware-malware normalized kernel values average of weighted API call graph for the different dataset sizes

	50	*100*	*150*	*200*	*250*	*300*
Benign—Benign	30.78	34.10	30.37	30.50	29.36	**31.41**
Malware—Malware	**31.43**	28.20	25.52	27.20	24.26	23.54

5.4.2.3 Benign-Malware Kernel Results

Considering that the kernel values are a similarity metric between two graphs, we present in Fig. 5.3 and Table 5.5 the kernel values from the comparison of benign and malware samples. For the largest dataset of 300 benign and malware samples, the average similarity of graphs is 11.15% and 12.21% for the unweighted and weighted graph approach, respectively. This clearly demonstrates that the API call graphs are a very effective and a significant feature for malware detection.

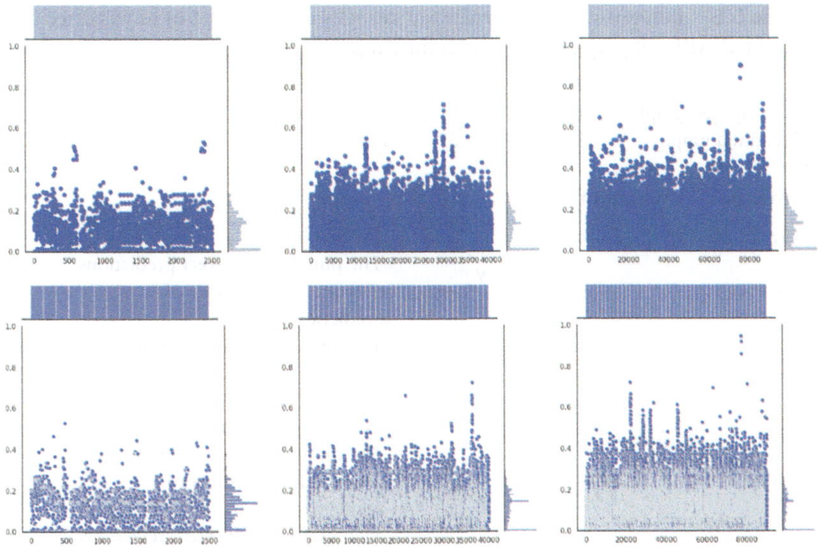

Fig. 5.3 Benign-malware kernel values for dataset size: (from left to right) 50, 200, and 300, for unweighted graph (top) and weighted graph (bottom)

5.4.3 SVM Training and Classification

We test the trained SVM classifier by using nine different train-to-test set splits, by randomly splitting the kernel values in a 10%–90% split and arriving to 90%–10% train-to-test split in 10% intervals. The whole procedure is repeated 100 times, and the resulting evaluation metrics are the average metric values over the 100 repetitions. The kernel values are split in groups containing an equal number of malware and benign originating kernel values: sets of 50, 100, 150, 200, 250, and 300. We evaluate the classification performance of the proposed framework using the accuracy, precision, recall, and the ROC curve of the SVM testing for the different train-to-test-splits (these evaluation metrics are elaborated upon in Table 5.6).

In Tables 5.7 and 5.8, we present the experimental results of the malware detection scheme for unweighted and weighted API call graphs, respectively. The unweighted approach achieves a maximum accuracy of 98.62% for the largest dataset, which is comparable to the results attained

Table 5.5 Benign-malware normalized kernel values average of unweighted and weighted API call graph for the different dataset sizes

	50	*100*	*150*	*200*	*250*	*300*
Unweighted graph	10.75	11.07	13.03	11.21	13.84	11.15
Weighted graph	13.37	12.32	11.63	12.35	11.93	12.21

Table 5.6 Evaluation metrics

Metric	*Equation*[a]	*Definition*
Accuracy	$\frac{TP+TN}{TP+TN+FP+FN}$	The number of correct predictions over the total number of predictions.
Precision	$\frac{TP}{TP+FP}$	The ratio of positive identifications that were actually correct.
Recall	$\frac{TP}{TP+FN}$	The ratio of actual positives that were correctly identified.
TPR	$\frac{TP}{TP+FN}$	The number of correct predictions over the total number of predictions
FPR	$\frac{FP}{FP+TN}$	The number of correct predictions over the total number of predictions
ROC Curve-AUC	-	A Receiver operating characteristic (ROC) curve shows the performance of a classifier at all classification thresholds. The ROC curve plots the TPR against the FPR at different classification thresholds. AUC measures the accumulated performance across all thresholds of classification. The ROC-AUC measures the quality of the classifier, no matter what classification threshold is chosen.

[a] TP is the number of True Positives, TN is the number of True Negatives, FP is the number of False Positives, and FN is the number of False Negatives

by similar attempts [21]; however, the dataset size we use is significantly smaller, and the computational cost is substantially reduced. On the other hand, the weighted graph approach surpasses 99.1% and almost reaches perfect accuracy while still functioning with a small dataset and keeping computational cost and operation time low.

Table 5.7 SVM evaluation metrics for unweighted graph

Benign-Malware	*Accuracy*	*Precision*	*Recall*	*ROC-AUC*
50–50	89.43	90.50	88.55	89.58
100–100	98.55	97.22	**99.93**	98.57
150–150	97.43	97.59	97.57	97.48
200–200	97.05	96.09	98.82	97.07
250–250	98.14	**98.19**	98.12	98.16
300–300	**98.62**	98.03	99.26	**98.59**

Table 5.8 SVM evaluation metrics for weighted graph

Benign-Malware	*Accuracy*	*Precision*	*Recall*	*ROC-AUC*
50–50	90.58	90.50	88.56	89.58
100–100	94.73	97.22	**99.92**	98.57
150–150	96.33	97.59	97.57	97.48
200–200	97.47	96.09	98.83	97.07
250–250	98.99	98.03	98.12	98.16
300–300	**99.15**	**98.88**	99.26	**98.58**

It is worth mentioning that the SVM algorithm is fine-tuned so as to avoid over- or underfitting and the dataset used has been examined and evaluated as per its good standing, and thus the detection results are reliable and robust. In Fig. 5.4, we present a comparison of our unweighted and weighted graph approaches.

As shown in Tables 5.7 and 5.8, the best classification accuracy is achieved by employing the SVM algorithm for weighted graphs. In Table 5.9, we compare the accuracy of our malware classification scheme against the accuracy achieved by similar state-of-the-art malware classification schemes. We note that one of the compared approaches employs an API call graph-based malware classification scheme which is relatively analogous to our method [21]. However, as mentioned before, in contrast to this conventional approach, we efficiently exploit the constraints of the problem and achieve higher accuracy faster and with a smaller dataset while also considering the case of weighted graphs. We also compare our classification scheme with other state-of-the-art malware classification approaches that employ API calls and/or graph-based solutions.

The comparison with state-of-the-art methods that employ API calls as features for malware classification demonstrates that our approach achieves higher accuracy than all static analysis approaches. Furthermore,

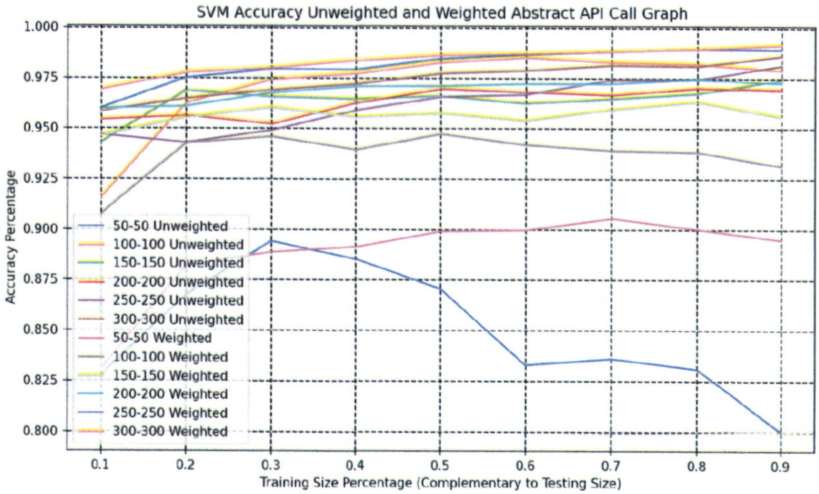

Fig. 5.4 SVM accuracy for unweighted and weighted abstract API call graph for the different dataset sizes and train-to-test splits

Table 5.9 Comparison with state-of-the-art malware classification approaches

Approach	*Accuracy*	*Notes*
[21]	98.91	Similar effort employing API calls on a Random Walk Graph Kernel with unweighted graphs.
[9]	96.50	Obfuscated malware detection using API call features and SVM to find optimal n-gram model
[11]	98.50	Extract API calls from disassembled executable and classify based on frequency of API Call usage
[52]	98.31	API call categorization based on appearance in benign or malicious executables
[46]	98.86	API call graphs transformed into low dimension numeric vector feature set introduced to DNNs
[71]	93.67	LSTM language model calculates similarity score based on Android system call sequences of executables
[22]	96.50	Android malware detection using features based on system calls (MALINE)
[42]	69.00	CNN applied to raw opcode sequences
[48]	94.00	Deep learning approach using the raw bytes of an executable file as input (MalConv)
[26]	98.30	HIN-embedding model metagraph2vec representing relatedness over files
Our Approach	**99.15**	Efficient API call graph classification

we observe that our proposed classification scheme outperforms both n-gram and language-based models, as well as graph-based approaches.

5.5 Conclusions

We introduced an efficient and effective static method for malware detection, which employs API call graphs. Our method is based on the calculation of an appropriate abstract API call graph, with reduced size taking into account problem constraints. Furthermore, it includes efficient calculation of a random walk graph kernel as a similarity metric. Through experiments using an appropriate dataset, we show that the calculated kernel constitutes an effective metric, which can be readily used for malware classification with machine learning methodologies such as SVM. Employing SVM and considering two different cases for the abstract API call graph, an unweighted and a weighted one, we demonstrate that our method is comparable to available alternatives when using unweighted graphs, reaching more than 99% accuracy, and outperforms alternatives when employing weighted graphs.

References

1. (2019). https://ghidra-sre.org/ [Online; accessed 12-July-2022]
2. (2022). https://virusshare.com/, [Online; accessed 12-July-2022]
3. (2022). https://www.microsoft.com/en-us/windows [Online; accessed 12-July-2022]
4. (2022). https://git-scm.com/ [Online; accessed 12-July-2022]
5. (2022). https://www.virustotal.com/gui/home/upload [Online; accessed 12-July-2022]
6. (2022). https://www.cygwin.com/ [Online; accessed 12-July-2022]
7. (2022). https://www.codeblocks.org/ [Online; accessed 12-July-2022]
8. Ah-Pine J (2010) Normalized kernels as similarity indices, pp 362–373. https://doi.org/10.1007/978-3-642-13672-6_36
9. Alazab M, Layton R, Venkataraman S, Watters P (2010) Malware detection based on structural and behavioural features of api calls
10. Alazab M, Venkataraman S, Watters P (2010) Towards understanding malware behaviour by the extraction of api calls. In: 2010

second cybercrime and trustworthy computing workshop. IEEE, New York, pp 52–59
11. Alazab M, Venkatraman S, Watters P, Alazab M, et al (2010) Zero-day malware detection based on supervised learning algorithms of API call signatures. AusDM 11:171–182
12. Alazab M, Alazab M, Shalaginov A, Mesleh A, Awajan A (2020) Intelligent mobile malware detection using permission requests and API calls. Futur Gener Comput Syst 107:509–521
13. Amer E, Zelinka I (2020) A dynamic windows malware detection and prediction method based on contextual understanding of API call sequence. Comput Secur 92:101760
14. Anderson HS, Roth P (2018) EMBER: an open dataset for training static PE malware machine learning models. ArXiv e-prints 1804.04637
15. Avrachenkov K, Chebotarev P, Rubanov D (2017) Kernels on graphs as proximity measures, vol 10519, pp 27–41. https://doi.org/10.1007/978-3-319-67810-8_3
16. Cai M, Jiang Y, Gao C, Li H, Yuan W (2021) Learning features from enhanced function call graphs for android malware detection. Neurocomputing 423:301–307
17. Canali D, Lanzi A, Balzarotti D, Kruegel C, Christodorescu M, Kirda E (2012) A quantitative study of accuracy in system call-based malware detection. In: Proceedings of the 2012 international symposium on software testing and analysis, pp 122–132
18. Chen ZG, Kang HS, Yin SN, Kim SR (2017) Automatic ransomware detection and analysis based on dynamic API calls flow graph. In: Proceedings of the International Conference on Research in Adaptive and Convergent Systems, pp 196–201
19. Cho IK, Kim T, Shim YJ, Park H, Choi B, Im EG (2014) Malware similarity analysis using api sequence alignments. J Internet Serv Inf Secur 4(4):103–114
20. Cortes C, Vapnik V (1995) Support-vector networks. Mach Learn 20(3):273–297
21. Dam KHT, Touili T (2017) Malware detection based on graph classification. In: Proceedings of the 3rd international conference on information systems security and privacy. SCITEPRESS—Science and Technology Publications. https://doi.org/10.5220/0006209504550463

22. Dimjašević M, Atzeni S, Ugrina I, Rakamaric Z (2016) Evaluation of android malware detection based on system calls. In: Proceedings of the 2016 ACM on international workshop on security and privacy analytics, pp 1–8
23. Ding Y, Zhu S, Xia X (2016) Android malware detection method based on function call graphs. In: Neural information processing. Springer International Publishing, Berlin, pp 70–77. https://doi.org/10.1007/978-3-319-46681-1_9
24. Ducau FN, Rudd EM, Heppner TM, Long A, Berlin K (2020) Automatic malware description via attribute tagging and similarity embedding. arXiv preprint arXiv:1905.06262
25. Elkhawas AI, Abdelbaki N (2018) Malware detection using opcode trigram sequence with SVM. In: 2018 26th International conference on software, telecommunications and computer networks (SoftCOM). IEEE, New York, pp 1–6
26. Fan Y, Hou S, Zhang Y, Ye Y, Abdulhayoglu M (2018) Gotcha-sly malware! scorpion a metagraph2vec based malware detection system. In: Proceedings of the 24th ACM SIGKDD international conference on knowledge discovery and data mining, pp 253–262
27. Gandotra E, Bansal D, Sofat S (2014) Malware analysis and classification: a survey. J Inf Secur 5:56–64. https://doi.org/10.4236/jis.2014.52006
28. Gärtner T, Flach P, Wrobel S (2003) On graph kernels: hardness results and efficient alternatives, vol 129–143, pp 129–143. https://doi.org/10.1007/978-3-540-45167-9_11
29. Harang R, Rudd EM (2020) Sorel-20M: a large scale benchmark dataset for malicious PE detection. arXiv preprint arXiv:2012.07634
30. Hassen M, Chan PK (2017) Scalable function call graph-based malware classification. In: Proceedings of the seventh ACM on conference on data and application security and privacy, pp 239–248
31. Hofmann T, Schölkopf B, Smola A (2007) Kernel methods in machine learning. Ann Stat 36. https://doi.org/10.1214/009053607000000677
32. Hou S, Saas A, Chen L, Ye Y (2016) Deep4maldroid: a deep learning framework for android malware detection based on linux

kernel system call graphs. In: 2016 IEEE/WIC/ACM international conference on Web Intelligence Workshops (WIW). IEEE, New York, pp 104–111
33. Imrich W, Klavžar S, Hammack RH (2000) Product graphs: structure and recognition. Wiley, New York
34. Jerlin MA, Marimuthu K (2018) A new malware detection system using machine learning techniques for api call sequences. Journal of Applied Security Research 13(1):45–62
35. Jiang H, Turki T, Wang JT (2018) Dlgraph: malware detection using deep learning and graph embedding. In: 2018 17th IEEE international conference on machine learning and applications (ICMLA). IEEE, New York, pp 1029–1033
36. Kang U, Tong H, Sun J (2012) Fast random walk graph kernel. In: Proceedings of the 2012 SIAM international conference on data mining. SIAM, New York, pp 828–838
37. Kelly Bissel PDC, LaSalle R (2019) Ninth annual cost of cybercrime study: the cost of cybercrime. Ponemon Institue LLC, Accenture plc
38. Ki Y, Kim E, Kim HK (2015) A novel approach to detect malware based on api call sequence analysis. Int J Distrib Sens Netw 11(6):659101
39. Kriege NM, Johansson FD, Morris C (2020) A survey on graph kernels. Appl Network Sci 5(1):1–42. https://doi.org/10.1007/s41109-019-0195-3
40. Kumar S, Singh CBB (2018) A zero-day resistant malware detection method for securing cloud using svm and sandboxing techniques. In: 2018 Second International Conference on Inventive Communication and Computational Technologies (ICICCT). IEEE, New York, pp 1397–1402
41. Ma Z, Ge H, Liu Y, Zhao M, Ma J (2019) A combination method for android malware detection based on control flow graphs and machine learning algorithms. IEEE Access 7:21235–21245
42. McLaughlin N, Martinez del Rincon J, Kang B, Yerima S, Miller P, Sezer S, Safaei Y, Trickel E, Zhao Z, Doupé A, et al (2017) Deep android malware detection. In: Proceedings of the seventh ACM on conference on data and application security and privacy, pp 301–308

43. Merabet HE, Hajraoui A (2019) A survey of malware detection techniques based on machine learning. Int J Adv Comput Sci Appl 10(1). https://doi.org/10.14569/ijacsa.2019.0100148
44. Nikolopoulos SD, Polenakis I (2017) A graph-based model for malware detection and classification using system-call groups. J Comput Virol Hacking Tech 13(1):29–46
45. Peiravian N, Zhu X (2013) Machine learning for android malware detection using permission and api calls. In: 2013 IEEE 25th international conference on tools with artificial intelligence. IEEE, New York, pp 300–305
46. Pektaş A, Acarman T (2020) Deep learning for effective android malware detection using api call graph embeddings. Soft Comput 24(2):1027–1043
47. Pluskal O (2015) Behavioural malware detection using efficient SVM implementation. In: Proceedings of the 2015 conference on research in adaptive and convergent systems, pp 296–301
48. Raff E, Barker J, Sylvester J, Brandon R, Catanzaro B, Nicholas CK (2018) Malware detection by eating a whole EXE. In: Workshops at the thirty-second AAAI conference on artificial intelligence
49. Ronen R, Radu M, Feuerstein C, Yom-Tov E, Ahmadi M (2018) Microsoft malware classification challenge. https://doi.org/10.48550/ARXIV.1802.10135. https://arxiv.org/abs/1802.10135
50. Salehi Z, Ghiasi M, Sami A (2012) A miner for malware detection based on API function calls and their arguments. In: The 16th CSI international symposium on artificial intelligence and signal processing (AISP 2012). IEEE, New York, pp 563–568
51. Salehi Z, Sami A, Ghiasi M (2014) Using feature generation from API calls for malware detection. Computer Fraud and Security 2014(9):9–18
52. Sami A, Yadegari B, Rahimi H, Peiravian N, Hashemi S, Hamze A (2010) Malware detection based on mining api calls. In: Proceedings of the 2010 ACM symposium on applied computing, pp 1020–1025
53. Schölkopf B (2000) The kernel trick for distances, vol 13, pp 301–307
54. Schölkopf B, Smola AJ, Bach F, et al (2002) Learning with kernels: support vector machines, regularization, optimization, and beyond. MIT Press, New York

55. Singh T, Di Troia F, Corrado VA, Austin TH, Stamp M (2016) Support vector machines and malware detection. J Comput Virol Hacking Tech 12(4):203–212
56. Souri A, Hosseini R (2018) A state-of-the-art survey of malware detection approaches using data mining techniques. Hum-centric Comput Inf Sci 8(1). https://doi.org/10.1186/s13673-018-0125-x
57. Sugiyama M, Borgwardt K (2015) Halting in random walk kernels. In: NIPS
58. Sundarkumar GG, Ravi V, Nwogu I, Govindaraju V (2015) Malware detection via API calls, topic models and machine learning. In: 2015 IEEE International Conference on Automation Science and Engineering (CASE). IEEE, New York, pp 1212–1217
59. Takeuchi Y, Sakai K, Fukumoto S (2018) Detecting ransomware using support vector machines. In: Proceedings of the 47th international conference on parallel processing companion, pp 1–6
60. Thomas R (2017) Lief—library to instrument executable formats. https://lief.quarkslab.com/
61. Tian R, Islam R, Batten L, Versteeg S (2010) Differentiating malware from cleanware using behavioural analysis. In: 2010 5th international conference on malicious and unwanted software. IEEE, New York, pp 23–30
62. Uppal D, Sinha R, Mehra V, Jain V (2014) Exploring behavioral aspects of api calls for malware identification and categorization. In: 2014 International conference on computational intelligence and communication networks. IEEE, New York, pp 824–828
63. Uppal D, Sinha R, Mehra V, Jain V (2014) Malware detection and classification based on extraction of API sequences. In: 2014 International conference on advances in computing, communications and informatics (ICACCI). IEEE, New York, pp 2337–2342
64. Verizon (2020) Data breach investigations report 2020. https://enterprise.verizon.com/resources/reports/dbir [Online; accessed 12-July-2022]
65. Vishwanathan S, Borgwardt KM, Schraudolph NN, et al (2006) Fast computation of graph kernels. In: NIPS, vol 19, pp 131–138
66. Vishwanathan SVN, Borgwardt KM, Kondor IR, Schraudolph NN (2008) Graph kernels. CoRR abs/0807.0093. http://arxiv.org/abs/0807.0093, 0807.0093

67. Wang T, Xu N (2017) Malware variants detection based on opcode image recognition in small training set. In: 2017 IEEE 2nd international conference on cloud computing and big data analysis (ICCCBDA). IEEE, New York, pp 328–332
68. Wu DJ, Mao CH, Wei TE, Lee HM, Wu KP (2012) Droidmat: android malware detection through manifest and API calls tracing. In: 2012 Seventh Asia joint conference on information security. IEEE, New York, pp 62–69
69. Wu WC, Hung SH (2014) Droiddolphin: a dynamic android malware detection framework using big data and machine learning. In: Proceedings of the 2014 conference on research in adaptive and convergent systems, pp 247–252
70. Xiao F, Lin Z, Sun Y, Ma Y (2019) Malware detection based on deep learning of behavior graphs. Math Probl Eng 2019:1–10
71. Xiao X, Zhang S, Mercaldo F, Hu G, Sangaiah AK (2019) Android malware detection based on system call sequences and LSTM. Multimed Tools Appl 78(4):3979–3999
72. Xiaofeng L, Xiao Z, Fangshuo J, Shengwei Y, Jing S (2018) ASSCA: API based sequence and statistics features combined malware detection architecture. Procedia Comput Sci 129:248–256
73. Ye Y, Hou S, Chen L, Lei J, Wan W, Wang J, Xiong Q, Shao F (2019) Out-of-sample node representation learning for heterogeneous graph in real-time android malware detection. In: Proceedings of the twenty-eighth international joint conference on artificial intelligence, international joint conferences on artificial intelligence organization. https://doi.org/10.24963/ijcai.2019/576
74. Yeo M, Koo Y, Yoon Y, Hwang T, Ryu J, Song J, Park C (2018) Flow-based malware detection using convolutional neural network. In: 2018 International conference on information networking (ICOIN). IEEE, New York, pp 910–913

CHAPTER 6

Deep Learning for Windows Malware Analysis

Mohamed Belaoued, Abdelouahid Derhab, Nassira Chekkai, Chikh Ramdane, Noureddine Seddari, Abdelghani Bouras, and Zahia Guessoum

6.1 Introduction

The emergence of the Internet has provided a powerful means of communication and data sharing, which has a huge impact on the worldwide economic growth. However, systems and networks have become exposed

M. Belaoued • N. Chekkai
Caplogy, Velizy-Villacoublay, France
e-mail: m.belaoued@caplogy.com; n.chekkai@caplogy.com

A. Derhab (✉)
Center of Excellence in Information Assurance (CoEIA), King Saud University, Riyadh, Saudi Arabia
e-mail: abderhab@ksu.edu.sa

C. Ramdane
LICUS, University of 20 Aout 1955 Skikda, Skikda, Algeria
e-mail: r.chikh@univ-skikda.dz

I. Almomani et al. (eds.), *Cyber Malware*, Security Informatics and Law Enforcement, https://doi.org/10.1007/978-3-031-34969-0_6

to different types of cyberattacks, which are launched by cybercriminals (i.e., Hackers) and could cause significant economic losses.[1] Malware is the suitable tool for hackers to launch cyberattacks. The term malware is used to refer to any computer program that was developed in order to perform malicious activities on computer systems. In the last 2 years, with the COVID-19 pandemic, we have witnessed a very concerning and alarming proliferation of Malware, with hundreds of thousands of malware samples that are discovered every day.[2] Moreover, the number of ransomware has doubled during the same period.[3] Therefore, deploying a robust anti-malware solution is vital in order to deal with malware proliferation.

The signature-based malware detection techniques, which have been widely used by the anti-virus vendors, are inefficient at detecting zero-day malware. Therefore, malware analysts have shifted to machine learning techniques in order to build intelligent and robust malware detection systems [77], which are more effective compared to the signature-based ones. However, they are facing several challenges. First, ML-based malware detection systems require a pre-processing phase called feature engineering, which aims at extracting features that characterize the analyzed malware samples, and are used as inputs to train the detection or classification model. Since these features are manually crafted by security researchers, and since most of existing malware instances are obfuscated using different

[1] https://cybersecurityventures.com/cybercrime-damages-6-trillion-by-2021/.

[2] https://www.comparitech.com/antivirus/malware-statistics-facts/.

[3] https://www.securitymagazine.com/articles/97166-ransomware-attacks-nearly-doubled-in-2021.

N. Seddari
LIRE Laboratory, Abdelhamid Mehri-Constantine 2 University, Constantine, Algeria
e-mail: noureddine.seddari@univ-constantine2.dz

A. Bouras
Department of Industrial Engineering, College of Engineering, Alfaisal University, Riyadh, Saudi Arabia
e-mail: abouras@alfaisal.edu

Z. Guessoum
CReSTIC EA 3804, University of Reims Champagne Ardenne, Reims, France
e-mail: zahia.guessoum@univ-reims.fr

techniques, such as packing, encryption, etc., the extraction of such characteristics can be a tedious task. Moreover, the high dimensionality that characterizes the extracted features requires a feature selection (reduction) phase that aims at removing irrelevant features, and which can be a labor-intensive task as well.

In the last decade, researchers have shifted to deep learning in order to overcome the aforementioned limitations of conventional machine learning approaches. Simply speaking, deep learning techniques can be defined as *as a neural network with a large number of parameters and layers* [70]. In fact, they are a subclass of machine learning algorithms that use many nonlinear information processing layers for supervised or unsupervised feature extraction, transformation, and classification, as well as pattern analysis [24]. Deep learning has widely been investigated in language processing [99] and have been extended to various fields, including cybersecurity [11, 12], and more specifically in the context of malware analysis and detection. In fact, deep learning has shown an impressive potential for detecting malware. Indeed, it can identify patterns in data that are too complex for traditional machine learning methods, making it more accurate and efficient. Additionally, deep learning can be used to detect new and unknown types of malware, making it an essential tool in the fight against cyberattacks. Indeed, employing deep learning techniques in the context of malware analysis and detection has various advantages, such as:

- Automatic feature learning from data is possible with deep learning, which can improve the detection accuracy.
- Deep learning can learn from data with different levels of abstraction, which can improve detection of more sophisticated malware.
- Deep learning can identify patterns that are too difficult for humans to discern.
- Deep learning can be used with unsupervised learning methods to detect previously unknown malware.

This survey paper aims at providing the most recent and comprehensive review of solutions that employ deep Learning for Windows malware analysis. The main contributions of this paper are the following:

- We provide a content-rich background about malware and malware analysis, as well as deep learning.

- We provide a detailed taxonomy that covers various classification criteria, namely, the analysis task, the type of extracted features, the feature representation method, and finally the deep learning algorithm.
- We comprehensively present the deep learning malware detection solutions for Windows malware and discuss them with regard to the proposed taxonomy.
- Furthermore, we provide an insight regarding the limitations and the challenges that face the existing deep learning malware analysis solutions, as well as some recommendations for future research.

This survey is organized as follows: Sect. 6.2 presents background concepts related to deep learning. Section 6.3 discusses the related surveys. In Sect. 6.4, we present the adopted research methodology. Section 6.5 presents our proposed taxonomy for deep learning malware analysis. In Sect. 6.6, we review, analyze, and discuss the current state-of-the-art malware detection solutions according to the proposed taxonomy. Section 6.7 highlights the open challenges for malware analysis using deep learning and recommends future research directions. Finally, Sect. 6.8 concludes the survey.

6.2 Deep Learning: Background and Basic Concepts

6.2.1 Definition

Nowadays, deep learning technique has attracted considerable attention because of its efficiency and usages. It can solve complex problems whose solutions did not exist before and even if they exist, they cannot achieve good results. Indeed, the arrival of deep learning has overcome several limitations of machine learning by dealing with high-dimensional data.

Deep learning (DL) is defined as a subfield of machine learning (ML) and artificial intelligence (AI) based on the use of multiple processing layers, in order to effectively extract useful features from the raw input, which can be used to handle multiple challenges in different application areas (e.g., natural language processing, cybersecurity, recommender systems, computer vision, healthcare, speech recognition, and many other fields) [55]. Deep learning algorithms imitate the human brain structure and function in order to compute information and then make decision to use multiple layers of neurons. In fact, as indicated in the following figure, DL

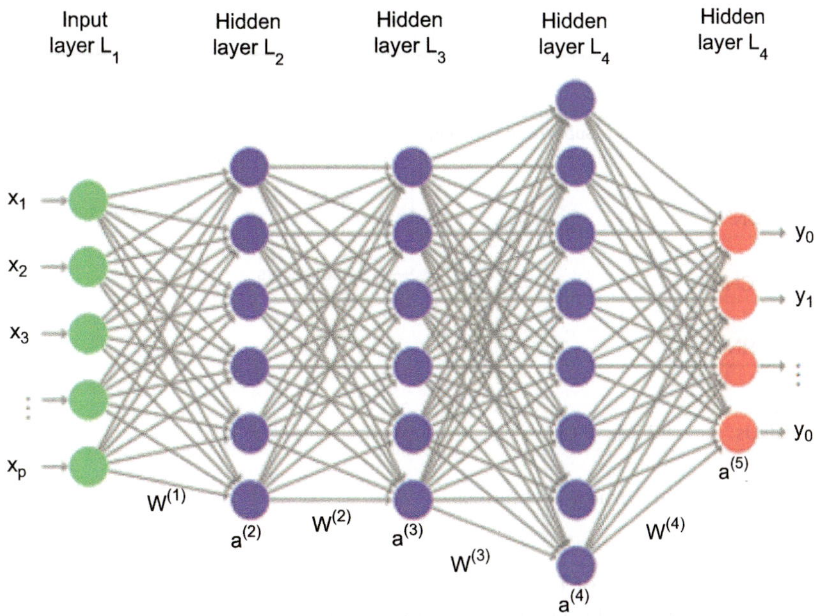

Fig. 6.1 General architecture of a deep neural network (source: towardsdatascience.com)

is made up of an input layer, an output layer, and one or more hidden layers, as shown in Fig. 6.1.

The **input layer** is made of one or more nodes that represent the artificial input neurons which receive the input data from the external environment. The inputs can, then, be normalized in ranges. The **hidden layer** is composed of one or many layers; it is responsible for the global processing of the network based on the data introduced by the input layer. Hidden layer uses activation functions in order to produce the results, by employing artificial neurons which calculate the weighted sum of the input data. Finally, the **output layer** is the last layer, and it provides the output data resulting from the processing of the hidden layer. More recently, a new category of neural networks has emerged, which are called graph neural networks (GNNs) [94]. GNNs are a powerful tool for learning on graph-structured data. GNNs learn to map node features to a low-dimensional representation and then use this representation to make predictions about

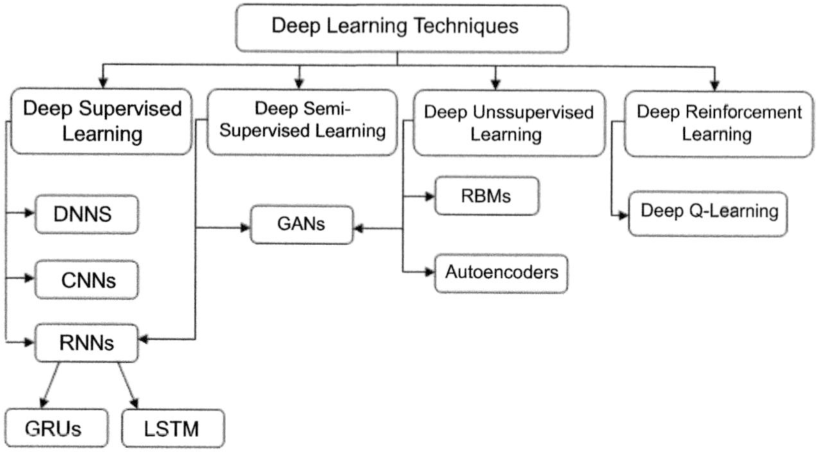

Fig. 6.2 Deep learning techniques

the graph structure or node labels. GNNs have been applied to a variety of tasks, including node classification, link prediction, and graph classification.

6.2.2 Deep Learning Techniques

Deep learning techniques are classified into four categories as shown in Fig. 6.2.

6.2.2.1 Deep Supervised Learning

This technique feeds labeled data. The sets of input and that of the resulting output are known and the deep learning model attempts to learn the mapping function. We find in this category various techniques, such as **deep neural networks (DNNs)** which are composed of three layers at least (input, output, and hidden), **convolutional neural networks (CNNs)** that are based on multiple multilayer perceptrons and hold one or many convolutional layers that are completely linked or pooled, and **recurrent neural networks (RNNs)** which successfully interpret temporal information. RNN class includes **gated recurrent units (GRUs)** technique which is based on particular memory elements that aim to quickly build recurrent neural networks by using few network parameters and **long short-term**

memory (LSTM) technique which is able to learn long-term dependencies in complex problem behaviors such as sequence prediction.

The key advantage of deep supervised learning is its simplicity and capacity of mapping data to produce known output results. Thus, we can identify the number of classes we want to have, and the obtained results are exact and credible. Hence, supervised learning algorithms were largely applied in forecast sales and risk evaluation. Nevertheless, the drawback of this technique is that the absence of some necessary samples can lead to exceeding the decision boundaries [2].

6.2.2.2 Deep Semi-supervised Learning

This technique feeds semi-labeled data including both tagged and untagged data which improves the learning performance. In this model, the outputs are known, but not all of them are labeled as the input. **Generative Adversarial Networks (GANs)** and **RNNs** are often employed as semi-supervised learning. One efficient application of semi-supervised learning is the text document classifier [35, 93]. The most important advantage of semi-supervised technique is the minimization of labeled data amount, whereas its disadvantage is the false output result that can be generated using inappropriate input features.

6.2.2.3 Deep Unsupervised Learning

This technique uses unlabeled data where only the inputs are known. Indeed, this model allows predicting the output results from incomplete and predefined labels. Clustering is included in deep unsupervised technique and was largely applied in many fields such as social networks analysis and anomaly detection. Among the most recent and efficient unsupervised learning techniques, we find **GANs, restricted Boltzmann machines (RBMs)** that use the connection between neurons of the same layer in addition to their connection with neurons of other layers, and **auto-encoders** which is capable of compressing and encoding data in an unsupervised way [16, 83]. Unsupervised learning algorithms allow discovering underlying patterns and effectively predicting relevant information. However, they are less accurate and computationally more complex compared to supervised learning.

6.2.2.4 *Deep Reinforcement Learning*

The deep reinforcement learning algorithms learn to act and react in an environment by using the most suitable actions. These algorithms use reactive agents in order to minimize the risk and maximize the reward. **Deep Q-learning** is a widely used reinforcement learning approach, which is based on the use of a deep neural network to learn the Q value of an action having a special state in its environment. The most advantage of deep reinforcement learning is its high performance of exploitation or exploration. It can also learn a set of action series. Thus, deep reinforcement learning algorithms find large applications in games and health areas. However, this technique requires a lot of computation due to the number of parameters [15].

6.2.3 *Deep Learning vs Machine Learning*

As mentioned before, deep learning is a subclass of machine learning; these two concepts are, hence, related to each other. The key differences between DL and ML are illustrated in Table 6.1.

The most important difference between deep learning and machine learning is the data dependency. As depicted in Table 6.1, deep learning algorithms require huge data to achieve good results, while machine learning algorithms can reach successful results with small data. Furthermore, ML needs structured data unlike DL that can work with both structured and unstructured data.

In terms of execution time, machine learning algorithms take only seconds or hours to train, while training a deep learning algorithm could take more than 2 weeks due to the large number of its parameters.

Table 6.1 Comparison between deep learning and machine learning

Feature	*Machine learning*	*Deep learning*
Data amount	Large	Small
Execution time (training)	Fast	Slow
Execution time (testing)	Slow	Fast
Data structure	Structured data	Structured and unstructured data
Hardware dependency	CPUs	CPUs and GPUs
Human intervention	Considerable	Little
Use	Simple and bi-complex problems	Complex problems

However, during the testing step, deep learning algorithms are faster than ML ones. Moreover, DL requires machines with significant computing power and multiple GPUs due to its big data, whereas ML can function on low-end machines with CPUs.

From Table 6.1, we can also observe that deep learning layers are able to learn and solve problems without human intervention, while machine learning model largely depends on the human intervention. Finally, machine learning is suitable for simple applications such as prediction and forecasting, while deep learning is used to solve complex problems.

6.3 Related Surveys

In the literature, there are several surveys on malware analysis. As shown in Table 6.2, we summarize the malware analysis surveys with respect to the following criteria:

- *Detection approach*: It indicates the malware detection approaches that are covered by the survey.
- *Operating system platform*: It indicates the targeted operating system platforms that are covered by the survey.
- *Outline and observations*: It states the main outline of the survey and any related observations.

From Table 6.2, we can observe that most of the surveys focused on two detection approaches: machine learning and deep learning. The surveys [45, 51, 58, 86] solely covered machine learning techniques. On the other hand, surveys in deep learning [72, 75, 81] only considered deep learning techniques. Other surveys covered both machine and deep learning techniques [32, 87]. In addition to machine learning techniques, Pan et al. [68] covered statistical detection models. Aslan et al. [5] presented detection techniques belonging to three detection approaches, including model checking, machine learning, and deep learning. Data mining approaches were covered in [82, 98], whereas malware analysis tools were presented in [26].

We can also observe that Windows and Android were the most investigated operating systems. The surveys [26, 32, 86, 98] and [45, 51, 58, 68, 72] only focused on Windows and Android OS, respectively. Other surveys considered the two operating system platforms [75, 81, 82, 87]. In addition to these two OS platforms, Aslan et al. [5] covered IoT malware.

Table 6.2 Related surveys on malware analysis

Reference	*Year*	*Detection approach*	*OS platform*	*Outline and observations*
Egele et al. [26]	2012	Malware analysis tools	Windows	Scope of the survey is restricted to dynamic analysis
Ye et al. [98]	2017	Data mining	Windows	Survey on malware detection using data mining techniques
Souri et al. [82]	2018	Data mining	Windows Android	Survey on malware detection using data mining techniques
Ucci et al. [86]	2019	Machine learning	Windows	Survey of machine learning techniques for malware analysis
Pan et al. [68]	2020	Statistical analysis	Android	Systematic literature review of Android malware detection
		Machine learning		Scope of the survey is restricted to static analysis
Aslan et al. [5]	2020	Model checking	Windows	Comprehensive review on malware detection approaches
		Machine learning	Android	
		Deep learning	IoT	
Liu et al. [58]	2020	Machine learning	Android	Review of android malware detection techniques using machine learning
Qiu et al. [72]	2020	Deep neural networks	Android	Survey of android malware detection using deep neural models
Sahin et al. [75]	2020	Deep learning	Windows Android	Survey on malware detection using deep learning techniques No taxonomy is provided Short survey
Gibert et al. [32]	2020	Machine learning	Windows	Systematic review of malware detection and classification techniques
		Deep learning		Taxonomy of features used by machine and deep learning
Urooj et al. [87]	2021	Machine learning	Windows	Scope of the survey is restricted to ransomware and to dynamic analysis
		Deep learning	Android	
Singh et al. [81]	2021	Deep learning	Windows	Survey on machine learning-based malware detection
			Android	No taxonomy is provided Short survey
Kouliaridis et al. [51]	2021	Machine learning	Android	Survey on machine learning techniques for android malware detection
Kambar et al. [45]	2022	Machine learning	Android	Survey on mobile malware detection techniques using machine learning
Our work		Deep learning	Windows	Survey on malware detection and classification using deep learning techniques

Some surveys are short [75, 81] or restricted to some analysis type like static analysis [68] or dynamic analysis [26]. Urooj et al. [87] only focused on ransomware using dynamic analysis. Four surveys [5, 32, 75, 81] are the closest to our work as they cover deep learning and target Windows operating system. However, our work differs from the four earlier-mentioned surveys in the following points:

- Multiple detection approaches and multiple operating system platforms are covered in [5]. In [75, 81], only deep learning approach is considered, and two operating systems, i.e., Windows and Android OS, are targeted. Differently, our work solely focuses on deep learning and only considers Windows OS.
- No taxonomy is provided in [75, 81]. In [32], feature-based taxonomy is proposed. Differently, our work proposes a taxonomy that classifies works with respect to different criteria.

6.4 Research Methodology

This section presents the adopted approach for selecting the reviewed papers, and which consists of data sources, search criteria, as well as the inclusion and exclusion criteria.

6.4.1 Data Sources and Search Criteria

In order to conduct a comprehensive literature coverage, and increasing the likelihood of finding high-quality papers, we selected a set of highly relevant databases, namely, ACM Digital Library, Springer Link, Science Direct, IEEE eXplore, PubMed, Web of Science, and Google Scholar.

We conducted an exhaustive search on the aforementioned databases using a search string that is based on various keywords such as "malware analysis," "malware detection," "deep learning," "Windows desktop," and "portable executable." The use search string is the following: "deep learning" and ("malware" and ("analysis" or "detection" or "classification") and ("Windows" or "desktop" or "portable executable" or "PE")).

6.4.2 Inclusion and Exclusion Criteria

In this survey paper, and in order to make the research more specific and comprehensive, we conducted a qualitative and quantitative study

Table 6.3 Inclusion and exclusion criteria

Criteria	*Description*
Inclusion	Papers that use deep learning techniques to detect malware
	Papers that provide solutions designed for Microsoft Windows desktop environment
	Papers that were peer-reviewed (if not, they must be highly cited)
Exclusion	Papers that use only traditional machine learning techniques
	Papers that are designed for other environments (i.e., Android, IoT, etc.)
	Papers that are not peer-reviewed (and with few citations) and short papers
	Papers written in a language other than English
	Papers that received less than five citations for the 3 years following their publication

of published research papers from 2015 onward by considering various inclusion and exclusion criteria, which are presented in Table 6.3.

6.5 Proposed Taxonomy

In this section, we introduce the proposed taxonomy (see Fig. 6.3) on how the existing solutions employ deep learning algorithms in malware analysis. In this taxonomy, we consider four main criteria for classifying the reviewed solutions, namely, the analysis task, the type of extracted features, the used feature representation method, and finally the used deep learning algorithm. These criteria are discussed in the rest of the section.

6.5.1 Malware Analysis Task

6.5.1.1 Detection

Malware detection is the process of identifying the presence of malware on a computer or network. Malware detection is a binary classification issue, where the outcome indicates whether the analyzed sample is malicious or benign. The detection task is generally the primary step in the malware analysis process. Indeed, once a sample is identified as malicious, it needs to be assigned to a specific category or family. This is the role of the classification task, and which is discussed in the following subsection. From the total number of surveyed papers, we observed that the majority of them (≈60%) propose malware classification solutions, while the rest of the papers deal with the malware detection problematic.

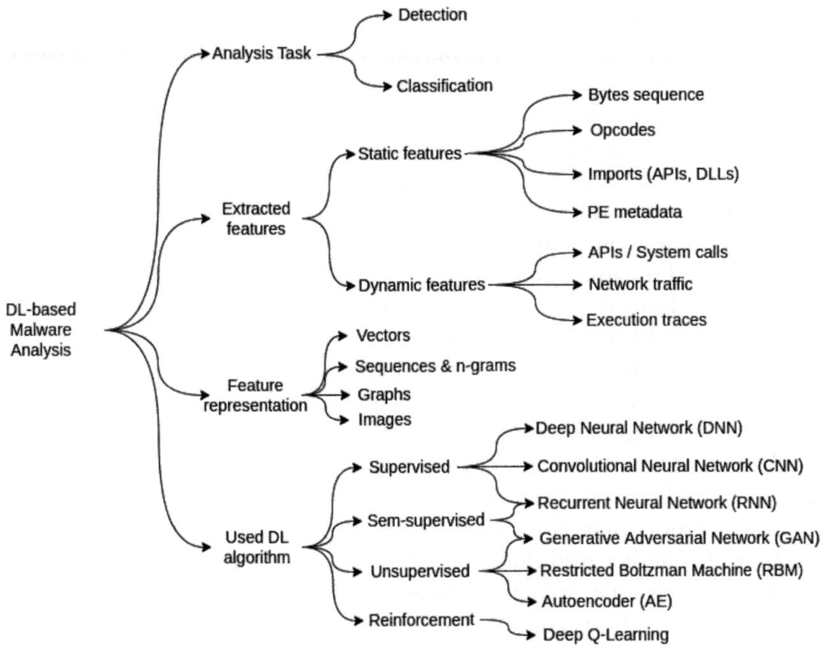

Fig. 6.3 Proposed taxonomy for malware analysis using deep learning

6.5.1.2 Classification

Malware classification is the process of categorizing malware according to its type or function. Indeed, malware classification is a more general approach to deal with malware. Rather than trying to detect every individual piece of malware, malware classification focuses on identifying and categorizing different types of malware. Indeed, it will assign the malware to a category or a family of malware with which it shares specific characteristics (e.g., target, behavior, etc.). This information can then be used to develop better detection and removal methods.

6.5.2 The Used Features

Malware detection or classification requires a preliminary stage, which is the analysis stage. The latter allows the extraction of the various attributes (features) that will be used to classify a file. There are two types of analyses, namely, static analysis and dynamic analysis [26, 80]. Dynamic analysis requires the execution of the program. This is carried out in a controlled setting that is typically created using an emulator (virtual environment) [26]. This can be useful for understanding how the malware interacts with the system and what is its purpose is. On the other hand, static analysis does not require the execution of the program; instead of that, the analyzed program is disassembled. The disassembly, also called reverse engineering, is the process of converting a compiled (machine code, bytecode) program into a more human readable format (i.e., assembly code). There are pros and cons to both static and dynamic malware analysis. For instance, static analysis is suitable for getting a general overview of what a piece of malware does (i.e., malicious or benign), but it can be limited in its ability to show how the malware actually behaves when executed and may not be able to uncover all of the malware's functionality since most existing malware are obfuscated. Dynamic analysis, on the other hand, is better for seeing how the malware behaves when executed, but it can be more difficult to set up and can be more time-consuming. Thus, these two types of analyses can be combined together resulting in what we call hybrid analysis. Based on the chosen analysis type, we can distinguish two main categories of features, namely, static features and dynamic features.

6.5.2.1 Static Features

Bytecode Data

A bytecode is a low-level language that is just like a hardware processor's assembly language (such as the IA-32 assembly language) [27]. It is often used to distribute programs or libraries in a platform-independent way. By examining the bytecode of a program, it is possible to identify previously seen patterns and thus efficiently determine if it contains a malicious code and preventing its execution on the system. Bytecode data has been widely investigated in the context of malware analysis and detection. The solutions presented in [3, 4, 18, 19, 21, 23, 42, 44, 48, 52, 54, 61, 64, 65, 88, 97, 100, 101] used bytecode data as main features, which represents ≈35% of the total number of the reviewed papers.

Operation Code

Operation code, also known as opcode, is a part of the assembly code instructions that identifies the operation to be performed (e.g., Push, Move, ADD, etc.) by the processor. Malware detection systems use

Fig. 6.4 A general PE file format structure

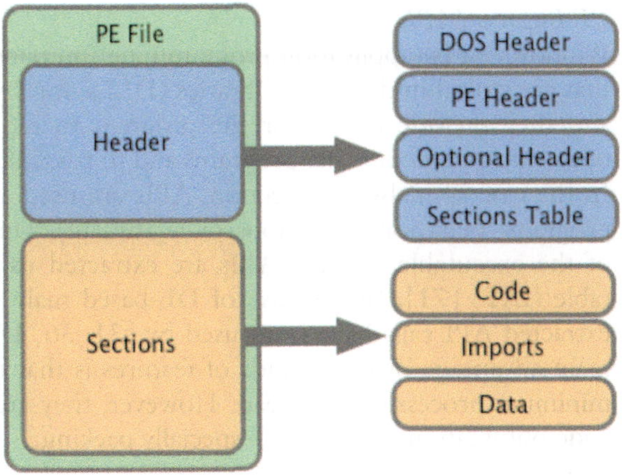

opcodes to identify malicious code. By analyzing the opcodes in a piece of code, a malware detection system can determine whether the code is malicious or not. In addition, by analyzing the opcodes, malware detection systems can identify which type of malware a piece of code is allowing to take the appropriate action. This is because different types of malware have different opcodes. For example, a piece of code that is designed to delete files will have different opcodes than a piece of code that is designed to steal information. The solutions presented in [21, 43, 46, 47, 50, 66, 67, 89, 97, 102, 103] employ opcodes as features, which represents ≈20% of the reviewed solutions.

PE Metadata

PE (portable executable) is the common file format for Microsoft Windows executable files [71]. A PE file is composed of several parts including headers (optional header, file header, etc.) as depicted in Fig. 6.4. The latter contains rich metadata regarding the file. PE metadata has been successfully leveraged in the context of malware analysis and detection, allowing the design of lightweight and highly accurate malware detection systems [7, 8, 78].

PE metadata has been also used to build deep learning-based malware detection systems, such as in [50, 73, 76, 90, 97].

PE Imports (APIs, DLLs)

Windows APIs (application programming interfaces) are a set of routines that are stored in dedicated libraries (DLLs, for Dynamic Link Libraries), and they provide a way for the program to request services from the operating system or other programs and to pass information back to them. In the case of malware detection, APIs are used as a mean to reflect the programs behaviors (malicious or legitimate). During the static analysis of the executable file, the APIs are extracted using the import address table (IAT) [71]. In the case of DL-based malware detection, statically extracted API calls have been used by [31, 36, 47, 50, 62, 76, 89]. The main advantage of these kinds of features is that they are extracted with minimum processing overhead. However, they are highly impacted with code obfuscation techniques, especially packing.

6.5.2.2 *Dynamic Features*

API/System Calls

APIs calls can be also extracted dynamically and that by running the program in a sandbox. APIs can request OS services through making system calls. System calls are a low-level way for a program to request a service from the kernel of the operating system it is running on. This may include requesting more memory, doing input/output, or creating a new process. In this case, API/system calls sequences are considered, since they provide a better description of the programs behavior, based on their chronological order. For instance, API/system call sequences have been employed by [1, 6, 22, 40, 49, 53, 56, 60, 61, 69, 85, 92, 95]. Moreover, the work of Huanfg at al. considers the API arguments as an additional indicator [41].

Execution Traces

By execution traces, we mean every action that is accomplished by the malware during its execution and that modifies the state of the system (i.e., host-based indicators). These can be file manipulations, registry updates, etc. The solutions presented in [22, 30, 85] employed such type of features.

Network Traffic

By analyzing network traffic, it is possible to identify malicious activities and take appropriate steps to mitigate the threat [9]. There is a variety of techniques that can be used for network traffic analysis, including packet inspection, flow analysis, and log analysis. Packet inspection is the most granular form of traffic analysis, as it allows analysts to examine each individual packet of data that passes through the network. Flow analysis groups together packets that are part of the same communication. Log analysis relies on data that has already been collected by network devices. The work of David et al. [22] and Shibahara et al. [79] employed network traffic features.

6.5.3 *Feature Representation Method*

There are various ways of representing malware features for detection purposes. Some common methods include using vectors, sequences, graphs, n-grams, and more recently image representation.

6.5.3.1 Vectors

There are many ways to represent features for deep learning, but one of the most popular is using a vector. A vector is simply a list of numbers, and each number in it represents a particular feature. For example, a vector might represent the set of features extracted from a malware sample. There are many benefits for using vector representations for deep learning. First, they are easy to work with and can be fed directly into most deep learning algorithms. Second, vectors are often able to capture complex relationships between features, which can be helpful for detecting patterns in data. Finally, vectors can be easily extended to include additional features, which can improve the accuracy of deep learning models. The solutions presented in [20, 31, 36, 46, 47, 49, 50, 61, 62, 79, 85] employed vector-based feature representation.

6.5.3.2 Sequences and n-Grams

There are many ways to represent sequence data for the purpose of malware analysis and detection. One common approach is to use a bag-of-words representation, where each instance is represented as a fixed-length vector of counts of words in the sequence. Another approach is to use a sliding window over the sequence, which results in what is called n-grams. An n-gram can be defined as ***an n-character slice of a longer string*** [17]. They are generated by shifting a windows of length l over that string. Formally, let S be the set of M^n distinct n-grams that can be formed from $\sum$. n-grams are all substrings of a larger string with a length of n [10]. As an example, from the word "M A L W A R E" we can extract the following 4-grams: "MALW," "ALWA,""LWAR," and "WARE." In the case of byte n-grams, the string represents the byte sequence of the analyzed file, and the byte n-grams are generated by shifting the window by n bytes. Opcode n-grams, similarly to byte n-grams, are generated by shifting a window on an opcode sequence. Ding and Siyi [25, 102] used that feature representation method by the conversion of the generated and which will be generated and concatenated to each other resulting in a unique opcode stream. Finally, n-gram features (with $n = 3$) are generated. For the reviewed solutions, those presented in [21, 31, 43, 49, 50, 56, 64, 73, 102] used sequence-based feature representation, while those presented in [1, 6, 30, 50, 53, 57, 60, 69, 84, 88, 92, 97] used n-gram-based feature representation.

6.5.3.3 *Graphs*

There are two main types of graphs that are frequently employed in malware analysis, namely, the control flow graph (CFG) and the function call graph (FCG). A control flow graph (CFG) is a graphical representation of the sequence of operations in a program. It is a directed connected graph, where each node represents an instruction of the file's assembly code and each edge represents an execution sequence [28]. A function call graph (FCG), on the other hand, is a graphical representation of the sequence of function calls in a program. Both CFGs and FCGs can be used to visualize the behavior of a program and to help debug it. However, they have different uses. CFGs are more useful for understanding the overall flow of a program, while FCGs are more useful for understanding the sequence of function calls.

In the case of DL-based malware detection, many researchers have opted for behavior graphs as the main feature representation for the analyzed samples. For instance, Ding and Siyi [25, 40, 102] opted for a CFG representation, while [95] opted for FCG one. In these pervious solutions, only the one of Hua et al. [40] used CFGs in its original from, since the latter solution employed a graph neural network, namely, deep graph convolutional network (DGCNN). For the rest of the solutions, they either transformed it into n-grams like the work of [102] or vector [95].

6.5.3.4 *Image Representation*

By representing images of malware, one can more easily find similarities and differences between samples. This helps to identify new malware and also to track the changes made to existing malware. There are various ways of representing images for this purpose, including 2D matrices, histograms, run-length encoding, and wavelets. In addition, images can be generated from the entire file(i.e., bytecode) or specific parts of it. Each of these approaches has benefits and drawbacks of its own, and the choice of representation will depend on the specific application and the used DL algorithms. Utilizing a convolutional neural network (CNN) is the most popular approach. CNNs can automatically extract features from images and learn to classify them. Other approaches include using a recurrent neural network (RNN) or a long short-term memory (LSTM) network, which can learn to detect patterns over time. Deep learning models can also be combined with traditional machine learning methods to improve the performance. There are a few potential advantages to this approach. For instance, it can be much faster than traditional scanning methods. In addition, it can be more accurate, since the entire file can be analyzed. The solutions presented in [3, 5, 18, 19, 21, 23, 38, 42, 44, 48, 52, 54, 65–67, 88–90, 97, 100, 101, 103] used image-based feature representation.

6.5.4 *Used DL Algorithms*

As presented in Sect. 6.2, there are many different DL algorithms that can be used for malware detection, but some of the most popular ones include convolutional neural networks (CNNs), which have been used by ≈55% of the reviewed solutions, and recurrent neural networks (RNNs), which have been used by ≈30% of the reviewed solutions. CNNs are often used for image classification tasks, while RNNs are better suited for sequence data. All of these algorithms have been used to build successful malware detection systems. However, there is no perfect algorithm that can be used for all tasks. It is important to choose the algorithm that is best suited for the specific problem at hand.

6.6 Description of Solutions

In this section, we discuss the reviewed solutions according to the proposed taxonomy. We also describe the solutions with respect to the size and nature (public or private) of the used datasets, the used performance evaluation metrics, the achieved results, as well as their weaknesses and strengths. The reviewed solutions have been grouped according the analysis task (i.e., detection or classification) as well as the extracted features (i.e., static and dynamic).

6.6.1 *Malware Detection Solutions*

6.6.1.1 Solutions that Employ Static Features

Saxe and Berlin [76] introduced a DL-based malware detection composed of three different components. The first component is the feature extractor, which allows extracting four types of features, namely, byte entropy histogram, string 2D histogram, PE imports, and PE metadata. The second one is the classifier, which is a four-layer deep feed forward neural network, with an input layer, two hidden layers, and an output layer. The input layer consists of 1024 nodes representing the input feature vector. The two hidden layers are also composed of 1024 nodes with a parametric rectified linear unit (PReLU) activation function [37]. The output layer predicts the output (malicious or benign) using the sigmoid function. The last component is the score calibrator, which represents the probability that a given file is malicious or not. Authors achieved 95% detection rate (DR) with 0.1% false-positive rate (FPR).

Hardy et al. [36] introduced DL4MD (deep learning framework for malware detection), which consists of two main modules, which are a feature extractor and a deep learning-based classifier. The feature extractor is used to extract API calls from the analyzed PE files using static analysis. The DL-based classifier is composed of several stacked auto-encoders (SAE). The authors evaluated DL4MD on a dataset containing 50K samples (45k training, 5K testing), and the results showed that the best configuration of the latter model (i.e., 3 hidden layers with 100 neurons in each layer) was able to achieve 95.46% of accuracy, which is 2% higher compared with the accuracy achieved by classical ML classifiers, namely, artificial neural network (ANN), support vector machine (SVM), Naıve Bayes (NB), and decision tree (DT).

Raff et al. [73] introduced an approach for efficient malware detection with minimum domain knowledge. Thus, the authors employ raw features(i.e., byte sequences), on which they will apply the minimum of pre-processing efforts. The authors introduced two baseline approaches as well as a deep neural networks based one. The first baseline approach uses PE metadata extracted using a third-party library called PortEX [34]. They extracted 115 features, which were fed to two machine learning algorithms, namely, random forest [14] and extra random trees [29]. The second baseline approach generates byte n-grams from the first 328 bytes of the PE file, which represent the location of the PE headers. These features are then fed to logistic regression algorithm. The last approach consists of using two types of neural networks, namely, fully connected neural networks (FCN) and recurrent neural network (RNN), both fed with the aforementioned raw byte region (328 bytes). Experimental results showed that the FCN model outperforms the three others with regard to area under ROC curve (AUC) and the balanced accuracy (BAC).

Choi et al. [18] introduced a malware detection approach that is based on the grayscale image representation of malware and benign binaries. They proposed a 256×256 images to represent the binaries, meaning that they only consider the first 64KB of the analyzed binaries. These images are then fed to a CNN composed of three convolutional layers, each followed by a pooling layer, in addition to two fully connected layers. They evaluated the model on a dataset composed of 10,000 benign and 2000 malware samples and achieved an accuracy of 95.66%.

In [46], a lightweight deep convolutional neural network-based method for detecting windows malware (CNN) is proposed. The proposed system is composed of two main components, which are the instructions analyzer and the classifier. The first components aims at disassembling the analyzed binaries, extracting the set of opcodes, grouping them by functionalities, and, finally, mapping them as 2D array. The results on the experiments of the detection system, based on a dataset contains around 70,000 samples, show an overall accuracy of 95% with a promising 10 hours as a training time of the system with one convolutional layer.

The study in [97] introduced MalNet, a novel self-learner malware detection approach, which uses CNN and LSTM networks. MalNet has two stages; the first one aims at statically analyzing the binaries and generating three types of features, namely, the grayscale image representation of bytecode, the opcode sequences, and various PE metadata. The second one is the core process of MalNet, in which the CNN and LSTM networks learn, respectively, from the grayscale images and the opcode sequences. In addition, and in order to optimize the detection performance, the authors used a stacking ensemble that integrates the two networks' output alongside with the metadata features and outputs the final prediction result. The model was evaluated on more than 40,000 samples collected from online software providers and Microsoft. The evaluation to an interesting achievement of the level of accuracy for malware detection measured at 99.88%. It also reached 99.14% of TPR with FPR of 0.1%.

Ding and Siyi [102] proposed a malware detection system composed of three main modules: the PE parser, the feature extractor, and the decision module. The PE parser aims at statically extracting the opcodes sequence and generating a control flow graph (CFG) from the analyzed file using IDA Pro tool [39]. The CFG is then converted into an opcode running tree from which n-gram features (with $n = 3$) are generated. In order to keep only the most relevant n-gram features, the authors employed document frequency, information gain, as well auto-encoder as feature selectors. The decision module is DBN-based and is fed using the generated n-gram feature vectors and is composed of three hidden layers each containing 200 units and an output layer that is composed of two units one for each output label (malware, benign). The experimental results indicated that the DBN model surpassed various "baseline classifiers," namely, k-nearest neighbor (KNN), decision tree(DT), and support vector machines (SVM).

A system that identifies malware programs using convolutional neural networks (CNNs) built on the AlexNet, ResNet, and VGG16 bases was proposed by Davuluru et al. [23]. This visualization is implemented by converting bytecode into a 2D matrix then visualizing it as grayscale images, which are then normalized for classification purposes. The restrictions of static and dynamic analysis can be circumvented by this form of visualization because it doesn't require any code disassembly and is computationally cheap. According to validation results from BIG 2015, CNN is a good feature extractor and classification tool. In Table 6.4, we provide a summary of the discussed malware detection solutions that employ static features.

6.6.1.2 Solutions that Employ Dynamic Features

In [69], Pascanu et al. proposed a method for malware detection that combines RNNs with multilayer perceptron (MLP) and logistic regression (LR). The RNN, which is trained in an unsupervised manner on dynamically extracted API call sequences, aims at learning a language model for the analyzed samples and constructing their feature representations. The latter are fed to the MLP and the LR classifiers, which classify the samples into either malicious or benign.

The work of Shibahara et al. [79] focuses on optimizing the analysis process by determining when to suspend it. They rely on the analysis of the network communications generated by the analyzed program with

Table 6.4 Summary of malware detection solutions that employ static features

Ref.	*Year*	*Features used*	*Feature rep.*	*Dataset size*	*DL Algo.*	*Results*
[76]	2015	Byte/entropy, PE imp meta	Histogram	431,926	DNN	Acc: 95.6–96.85%
[36]	2016	API calls	1-hot vect	50k	SAEs	Acc: 95.64%
[18]	2017	Bytecode	Image	12k	CNN	Acc: 91%
[102]	2017	Opcode	CFG, n-grams	4600	DBN	Acc: ≈98%
[97]	2018	Bytecode, Opcodes, PE Metadata	Images, sequences	40k	CNN, LSTM	Acc: 99.88%
[46]	2018	Opcodes	2-D array	70k	CNN	Acc: 95%
[19]	2019	Bytecode	Grayscale image	5k	CNN	Acc: 91.9–97.6%
[23]	2019	Bytecode	Grayscale image	≈10,000	CNN	Acc: 99.4 91.9–97.6%

a command and control server (C&C). To achieve this, the authors considered different characteristics of malware communication (i.e., the modified communication's intent and its shared latent purpose). The authors employed recursive tensor neural network (RSTNN) to decide when to stop the analysis for each sample and were able to reduce the total time taken by 67% compared with a conventional analyses methods.

Tobiyama et al. [85] proposed an approach for host-based malicious activity detection by monitoring and analyzing the processes' behaviors. The processes are represented by their ID, the executed operation (API), its result (success, access denied, etc.), etc. These information are stored in log files and will be used as raw features to train the feature extractor module, which is recurrent neural network (RNN) with LSTM units. The RNN then outputs a features vector, which will be then converted into an image, which contains local features representing processes' activities. These images are then fed to a CNN, which is responsible for the classification of the processes into malicious or benign.

Athiwaratkun et al. [6] proposed two deep learning models for malware detection. In the first model, RNN with LSTM units and GRU (i.e., gated recurrent unit) were investigated in order to build the features associated with different API call traces. The latter features are then used to train a fully connected layer and logistic regression algorithm. The second model, which is a CNN with six convolutional layers and three fully connected one, is trained on character-level features of size 1024 representing various events. Experimental results show that the LSTM-GRU model achieved a higher accuracy than the CNN one.

Maniath et al. [60] proposed an approach for crypto-ransomware detection that consists of collecting three different dynamic features, namely, API calls, file operations, and registry values. The aim is to be able to capture crypto-ransomware behavior patterns (e.g., pre-encryption phase). For this purpose, long short-term memory (LSTM) algorithm is used in order to classify the API calls sequences. The proposed approach was able to achieve good accuracy rate; however, the analysis phase took 20 minutes to complete, which was the major limitation of this work.

In [61], a malware detection system based on hybrid features and deep neural network is introduced. The authors did not implement any file analysis step, and instead they used an existing dataset,[4] which contains both static (e.g., file sections, entropy, assembly n-grams, etc.) and dynamic (e.g., contacted IP, DNS queries, execution processes, AV signatures, etc.) features of four different malware families. The authors considered only two malware families and trained a deep neural network to detect these two families.

Xiao et al. [95] designed a new behavior-based deep learning framework called BDLF. Instead of using API call sequences, the goal of BDLF is to gain deeper semantics in behavior graphs (e.g., n-gram). The authors investigated a deep learning model based on stacked auto-encoders (SAEs) to automatically obtain high-level representations of malware behaviors. In the proposed framework, they first constructed behavior graphs using the extracted API calls. In order to extract high-level characteristics from behavior graphs, authors used SAEs. The findings of the experiment show that BDLF can extract more useful abstract features and increase the accuracy of malware detection. Yakang et al. [40] directly fed the extracted FCG to a deep graph convolutional network (DGCNN). Then they use an algorithm to strip the subgraph that represents the unpack function call in the function call graph of the malware's packet. Authors then run the expansion operation on the subgraph which only contains local function call graph to get control flow graph of the packed malware. The features

[4] https://marcoramilli.com/2016/12/16/malware-training-sets-a-machine-learning-dataset-for-everyone/.

of the control flow graph are extracted as the input of the DGCNN for training, and the classifier is obtained to detect the packed malware.

Ding and Zhu [102] focused on studying the following problems: (a) how to build a malware detection system based on DBNs, (b) whether the unlabeled data can be used to improve the accuracy of malware classification, and (c) whether the deep representation generated using DBNs is helpful for feature extraction and dimension reduction. Therefore, authors represented the malware program as opcode sequences and extract the opcode n-grams to specify the behavioral features of malware. The architecture of proposed system consists of three main components: the PE parser, the feature extractor, and the malware detection module. The testing results show that the proposed model has better classification results than other models: support vector machines, decision trees, and the k-nearest.

Darabian et al. [20] studied the potential of applying deep learning techniques to detect cryptomining malware by using both static and dynamic analysis approaches. They used long short-term memory (LSTN) and convolutional neural network (CNN) techniques to advance the analysis of cryptomining malware. They considered a set of hybrid features composed of the captured system call events and opcode sequences. The proposed system achieved an accuracy rate of 95% using static features and an accuracy rate of 99% using dynamic ones.

An effective approach based on deep learning analysis for malware detection and explanation is proposed by Wang et al. in [91]; they used a classifier to predict whether the sample is malicious and an interpreter to explain the classifier result via a system call number sequence of the target sample with instrumentation tools in an elaborated sandbox. The approach just needs a small amount of feature data and can reduce the input dimension of the training model. The authors also adopted the layer-wise relevance propagation (LRP) algorithm to save the malware analyst time and to find which slice of a sequence makes the greatest contribution in the decision.

Aditya et al. [1] introduced an approach for detecting malware based on deep neural network and utilized a API call sequences. The model is implemented with two different recurrent neural network architectures for comparison (LSTM and GRU). The classification model that has been created employs the LSTM architecture with RMSProp optimizer, and a learning rate parameter shows that LSTM is better than GRU, achieving an accuracy of 97.3%.

Table 6.5 Summary of malware detection solutions that employ dynamic features

Ref.	*Year*	*Features used*	*Feature rep.*	*Dataset size*	*DL Algo.*	*Results*
[69]	2015	API calls	Sequences	500K	RNN,ESN,MLP	TPR:71%, FPR:0.1%
[79]	2016	Network traffic	1-hot Vect	29,562	RSTNN	F-score: 96.9%
[85]	2016	Process behavior	1-hot vect, Image	26	RNN(LSTM), CNN	AUC: 96%
[6]	2017	System Calls	Sequence	75k	LSTM, MLP	Acc: 95.6%
[60]	2017	API	Sequence	157	LSTM	96.67%
[61]	2018	PE, Bytec, APIs, Net. traffic	Vector	3772	DNN	Acc: 97%
[95]	2019	Call API	graph	1760	SAE	Acc: 98.6%
[20]	2020	Opcodes, system calls	Scale values, binary vectors	1500	LSTN,ATT-LSTM,CNN	Acc: 95.99%
[40]	2020	Functions calls	Graph	600	DGCNN	Acc: 96.4%
[1]	2021	API calls	Sequences	2210	LSTM	Acc: 97.3%
[91]	2021	API calls	Sequences	2950	M-Bi-LSTM	Acc: 97.39%
[30]	2021	Execution traces	Sequences	4000	CNN + LSTM	Acc: 91.63%
[56]	2022	API calls	n-gram, sequences	43,007	CNN + LSTM	Acc: 97.31%

Ghanei et al. [30] presented a dynamic malware analysis method which utilizes hardware events as feature inputs to the classification model during programs' execution. They used hardware events-based features in three ways. Firstly, the feature set of each period are given to the convolutional neural network (CNN) separately. Secondly, a time series is formed and is given to a long short-term memory (LSTM) network. Thirdly, a fully connected network is used between the outputs of CNNs and the LSTM network to model a voting classifier. The results showed that the combination of hardware events with voting network can be effective and can reach 91.63% of accuracy. In Table 6.5, we provide a summary of the discussed malware detection solutions that employ dynamic features.

6.6.2 *Malware Classification Solutions*

6.6.2.1 Solutions that Employ Static Features

In [49], the authors proposed an approach for malware families identification using system calls sequences. The latter are extracted by dynamically analyzing malware and benign samples using Cuckoo Sandbox [33] and are used to construct the execution paths of the analyzed samples. The redundant sequences are then removed, and the remaining ones are represented as 1-hot vector of length 60 (i.e., the 60 documented system calls) representing the presence or not of a specific kernel API call. These feature vectors are then fed to a convolutional network as 3-grams (3x60). The CNN will act as a feature extractor and will generate two features vectors, which are forwarded to the RNN part of the neural network. API traces dependencies are then modeled, and mean-pooling approach is used to extract features of highest importance from the LSTM output, which are forwarded to the Softmax layer that classifies each instance into a family. The experimental results show that the combination of CNN-RNN achieved better performances than the two models separately as well as two machine learning classifiers, namely, hidden Markov model (HMM) and support vector machine (SVM).

Zhang et al. [103] introduced IRMD, which is a malware families classification method based on CNN and an image-based representation of opcode sequences. In the proposed method, the authors first disassemble binary executables and extract opcode sequences that are represented as 2-D array, which is then converted to grayscale images. The authors then applied image processing techniques on the generated images, such as histogram normalization, dilation, and erosion. The resulting images were then fed to a CNN. The latter has a baseline three-level architecture composed of a convolutional layer, a pooling layer, and a fully connected layer. A softmax function is then used to classify malware variant and benign images. IRMD was evaluated on a dataset collected from VxHeavens repository and composed of 9168 malware samples from 10 distinct malware families and 8640 benign samples and was able to achieve 96.7% of accuracy. Mourtaji et al. [65] also used a CNN with image representation of malware samples. They were able to achieve the highest accuracy (i.e., 99.88%) on two distinct experiment settings on Microsoft BIG15 dataset. Similarly, Kumari et al. [52] relied on image representation of the analyzed binaries and CNN for malware families classification. However, they introduced three different CNN architectures. The first one has baseline architecture that is composed of three convolutional layers. Each convolutional layer is followed by a max-pooling layer and a ReLU activation layer. The second one is based on the VGG-16 architecture which is pre-trained on the ImageNet dataset, which is composed of 1000 classes. In the last model, the authors fine-tuned the last convolutional block of the VGG-16 model as well as the top-level classifier. Yue et al.

[101] and Rahul et al. [74] also employed CNN for malware families classification using image representation. In the first work [101], they trained a very deep neural network (DNN) composed of ten layers and a complex pre-processing method on the MalImg dataset and achieved an accuracy of 97.32%, while in the second one [74], they trained a baseline CNN architecture, two convolutional and two dense layers on the BIG 2015 dataset Kalash et al. [44]. Hemalatha et al. [38] used a pretrained densely connected convolutional network (DenseNet) model with class-balanced loss function for reweighting the categorical cross-entropy loss in the final classification layer. The DenseNet model uses fewer parameters and ensures information flow by connecting all the layers in the network with their feature maps. The performance of the proposed model was evaluated on four malware datasets, namely, Malimg, BIG 2015, Malicia, and Malvis, achieving, respectively, 98.23%, 98.46%, 89.48%, and 98.21% of accuracy. Zhihua et al. [19] developed an approach to advance the detection of malicious programs using convolutional neural networks (CNNs) and non-dominated sorting genetic algorithm II (NSGA-II). The CNNs are used to identify and classify grayscale images converted from executable files of malicious code. NSGA-II is then employed to deal with the imbalanced data of malware families. A series of experiments are performed for malware image data from Vision Research Lab, and the results show that the proposed method is effective maintaining higher accuracy. Ni et al. [67] considered opcode sequences instead of bytecode. They encoded these sequences using SimHash, which they considered as pixels and converts them to grayscale images. Kebede et al. [48] opted for a deep learning architecture composed of multilayer neural network with auto-encoders applied on malware images. An approach based on visualization and fine-tuned CNN is proposed by Vasan et al. in [88]; they used color instead of grayscale images generated from the malware binaries to identify and detect both packed and unpacked malware. The proposed method is called image-based malware classification using ensemble of CNNs (IMCEC). According the experimental result on Malimg malware benchmark, the proposed model demonstrated 99% accuracy for unpacked malware and 98% accuracy for packed malware. The problem with this approach is that it considers the entire program's binary, which is very large and takes considerable time to process.

Venkatraman et al. [89] presented a hybrid model by employing similarity mining and deep learning architectures for accurately detecting and classifying obfuscated malware into their malware families. The proposed model used two types of learning approaches: CNN and LSTM. The objectives are (1) to describe the use of image-based techniques for identifying suspicious system behavior and (2) to suggest and research the use of hybrid image-based approaches with deep learning architectures for an efficient malware detection. The performance of the models is evaluated on the three datasets: VX Heavens, Malimg, and Microsoft. The model accuracy achieved 96% on average and the advantage that it required less computational cost as compared to the classical machine learning-based methods.

The work of [100] introduces MDMC, a byte-level malware classification approach based on Markov images and deep learning. In contrast to grayscale images, the first phase of MDMC does not take the issue of resizing into account and instead attempts to transform malware binaries into Markov pictures according to the bytes transfer probability matrix. The deep convolutional neural network is then used to classify Markov pictures. In this procedure, only malware binaries were employed; dynamic and reverse analysis were not used. On the Microsoft dataset and the Drebin dataset, two malware datasets, the performance of the suggested model has been assessed. On the two datasets, the average MDMC accuracy rates are 99.26% and 97.36%, respectively.

Lin and Yeh [57] presented an efficient one-dimensional convolutional neural network CNN models for malware classification. The 1D CNN models explore both bit-level and byte-level sequences extracted from malware executables. The authors designed a simple architecture of 1D CNN to learn the features from raw binary sequences and to convert malware executables into images. The experiments show that the proposed 1D CNN model achieves better performance with smaller resizing byte sequences with an accuracy of 96.32% and 98.70% using two benchmark datasets.

The comparison of the classification capabilities of convolutional neural networks (CNN) and extreme learning machines (ELM) for malware images classification is the main objective of Jain et al.'s [42] work. They used both two-dimensional images and one-dimensional vectors produced from images to view malware samples as images and apply image analysis algorithms. Results on the Malimg dataset showed that ELMs train faster than CNNs and produce results with higher accuracy while processing 1D data. The authors also noted that ELMs handle 2D data more quickly than CNNs. Finally, authors concluded that ELMs are faster to train than CNNs, but only by a relatively small factor as compared to image-based training.

The deep learning approach in [54] is practical for real-life uses since it has two interesting properties: it does not require neither feature engineering nor a long time to classify the malware class of a binary file. Indeed, the malware samples are converted into grayscale image representation and then fed to different neural network models. The latter are a combination of convolutional layers which process the input, with RNN and LSTM layers. The test conducted on the malware data from the Microsoft Malware Classification Challenge (i.e., BIG 2015) available on Kaggle (10,868 samples) shows an accuracy of 98.2% in the cross-validation procedure through the CNN bi-directional LSTM model.

In [66], authors proposed an ensemble learning-based classification system comprised of convolutional network to classify malware programs. In their research, they used the nine-class Microsoft Malware Classification Challenge (BIG 2015) dataset. For each malware file in this dataset, there is an assembly file and a compiled file. Convolutional neural networks are used to classify compiled files and display them as images; then convolutional neural networks (CNNs) are used to classify these images. Long short-term memory (LSTM) networks are used to classify machine language opcodes in assembly files after they have been converted into sequences. When identifying assembly files using an LSTM network, accuracy is 97.2 percent; when categorizing compiled files with a CNN architecture, accuracy is 99.4 percent.

Darem et al. [21] suggested a semi-supervised method for detecting obfuscated malware that combines opcode analysis, feature engineering, image processing, and deep learning approaches. The proposed approach transforms the malware binary into image for visual analysis of the malware executable and contrasts with well-known grayscale image-based classification methods. As a result, the approach identifies and predicts associated malware families with minimal running time overhead. They validated the proposed method through comprehensive experiments and compared it with other methods. Experimental results proved that the proposed approach achieved the highest performances with 99.12% of accuracy.

The work of [4] presents a new malware classification framework based on a hybrid deep learning algorithm. The framework combines two pretrained deep neural networks, namely, RestNet and Alexnet, in order to learn features from malware samples, which are represented as grayscale images, and classify them into different families. The framework is evaluated on a large dataset of malware samples and achieves state-of-the-art performance.

Mohammed et al. [64] introduced a malware detection mechanism based on convolutional neural networks (CNNs) and malware binaries images in the frequency domain ("bigram-dct" images). The authors proposed a joint feature metric that justifies the combination of two different features from byteplot images and "bigram-dct" images to create an accurate ensemble model for malware detection. They evaluated the proposed model on large dataset called MaleX consisting of Windows executable samples, obtaining an accuracy of 96%.

Meng et al. [62] introduced MCSMGS, which a malware classification model based on deep learning and statically extracted API calls sequences. The latter are considered malware gene sequences and are fed to a CNN. MCSMGS was evaluated on dataset composed of 5647 samples, obtained from VxHeavens repository, and it was able to achieve 98% of accuracy.

Kang et al. [47] introduced an approach for classifying malware into different categories using opcode and API sequences fetaures with word2vec and long short-term memory (LSTM) network. Moreover, authors showed the possibility that word2vec can be applied to classify malware and it can victorizes data using fewer dimensions than one-hot encoding. According to the experimental results on using the Microsoft Malware Classification dataset, the proposed model has 97.59% as accuracy classification rate.

Gibert et al. [31] have designed a novel multimodal deep learning framework named HYDRA, which is a network structure with multiple inputs and single output, for malware classification. The suggested approach combines end-to-end components with hand-engineered features in a modular architecture to categorize malware using CNN, DeepConv, and Malconv models. The features, which are learned via different network architectures, are fed through byte opcodes and API calls. Finally, all of the functions are connected seamlessly. Testing results show that the model achieved 99.75% of accuracy on the Microsoft BIG 2015 dataset.

In [3], authors investigated that most large datasets that include malicious and non-malicious programs are not public. To reduce this limitation, first they developed a new large public dataset for malware classification it called MC-dataset-multiclass (malware and clean ware in multiclass scenario). This dataset was then used to train a multiclass classification RNN, namely, an LSTM. Unknown programs were used to test this model for interpreting unstructured data. Evaluation findings indicate that the accuracy was 67.60%, with six classes containing five separate malware kinds.

A malware detection and family classification framework for malware based on deep neural networks and visualization is proposed by Jian et al. in [43]; they convert an executable file samples into asm files bytes files by disassembly technology. As a result, a balanced experimental dataset containing normal software samples and malware samples is constructed. To this end, the authors designed a new data representation approach based on the binaries and word vectors extracted from both asm files and bytes files and combined visualization technology with data augmentation to build an optimized deep neural network architecture, i.e., SERLA (SER esNet50 + Bi- L STM + A ttention) for malware detection. The experimental results show that proposed method is superior to the state-of-the-art methods and can achieve 98.31% accuracy. Li et al. [56] proposed a deep framework for malware detection using deep learning models, which is based on multiple API sequence intrinsic features. The proposed method is able to detect whether the software is malicious or not and to distinguish between malware and goodware. The authors firstly applied embedding and convolutional layers to well depict the actual software behaviors. Secondly, they designed an encoder to represent the semantic information of APIs and the relationship between API calls using the Bi-LSTM module. The experiments show that the proposed method performs better than all the baselines in using API sequence to detect the malware, achieving an accuracy score of 97.31%. In Table 6.6, we provide a summary of the discussed malware classification solutions that employ static features.

6.6.2.2 Solutions that Employ Dynamic Features

David and Netanyahu [22] introduced a malware classification approach based on behavior signature generation and deep belief networks (DBNs). The proposed approach uses a dynamic analysis technique to extract the behavior of each analyzed file using cuckoo sandbox. The resulted log file contains various information about the program's behavior, such as API calls, file manipulations, IP addresses, URLs, etc. The log file is then parsed and converted into a fixed-sized binary vectors that will be provided as an input to the DBN. The latter is composed of eight layers of auto-encoders, and the output layer contains 30 neurons. The authors experimented their approach on a dataset composed of 1800 samples and achieved an overall accuracy of 98.6%.

Table 6.6 Summary of malaware classification solutions that employ static features

Ref.	*Year*	*Features used*	*Feature rep.*	*Dataset size*	*DL Algo.*	*Results*
[103]	2016	Opcode	Image	17,808	CNN	Acc: 96.7%
[101]	2017	Bytecode	Image	9435	CNN	Acc: 97/32%
[52]	2017	Bytecode	Image	21,741	CNN	Acc: 97.07%
[48]	2017	Bytecode	Image	10,826	AE	99.15%
[62]	2017	API calls	Word2Vect	5647	CNN	Acc: 98%
[50]	2017	PE Meta, imp, opcod	n-gram, vector	22,757	FFNN, CNN	F1S: 92%
[54]	2018	Bytecode	Image	10,860	CNN + biLSTM	Acc: 98.2%
[44]	2018	Bytecode	Image	≈30k	CNN	Acc: 99.97%
[67]	2018	Opcode	Image	≈10k	CNN	Acc: 98.862%
[3]	2019	Bytecode	Image	19,740	CNN	Acc: 97.19%
[65]	2019	Bytecode	Grayscale image	≈30k	LSTM	Acc: 97.02–99.88%
[47]	2019	Opcode, API calls	Binary vectors	10,868	LSTM	Acc: 97.59%
[89]	2019	Opcode, API calls	Image	≈30k	CNN,LSTM	Acc: 96%
[73]	2017	PE metadata	n-gram	≈95K	CNN, RNN	Acc: 90.8–97.7%
[31]	2020	APIs, Bytec, Opcode	Binary vectors, n-gram	≈10K	Multimodal CNN	Acc: 99.75%
[88]	2020	Bytecode	Sequence, Image	≈10K	CNN	Acc: 98.99%
[66]	2020	Opcode	Image	≈10K	LSTM,CNN,RNN	Acc: 97.2, 99.4, 99.8%
[100]	2020	Bytecode	Image	≈15K	CNN	Acc: 99.26, 97.36%
[42]	2020	Bytecode	Image	9300	CNN,ELM	Acc: 96.3, 97.7%
[21]	2021	Opcode, bytecode	n-gram, image	10,868	CNN + XGBoost	Acc: 99.12%
[4]	2021	Bytecode	Image	≈40k	Alexnet, restnet	Acc: 97.78%
[43]	2021	Opcode	n-gram	10,868	CNN + RNN	Acc: 98.31%
[64]	2021	Bytecode	n-gram, image	179,725	CNN	Acc: 96.15%
[38]	2021	Bytecode	Image	21,741	DenseNet	Acc: 98.46%
[57]	2022	Bytecode	Image	10,868	CNN	Acc: 96.32%

Huang and Stokes [41] introduced MtNet (i.e., multi-task neural network), which is a system for malware detection and family classification. The proposed system provides a lightweight emulation engine that aims at extracting dynamic raw features, which are as follows: API calls, their input arguments, in addition to null terminated objects, which are the result of code unpacking process. During the preprocessing phase, which is a manual feature engineering processes, the API calls that have the same role are mapped into higher level concept, resulting in 114 API calls in total. The latter are combined with n-gram (with $n = 3$) features. In order to reduce the number of generated features, and only keep the most relevant ones, MtNet integrates a features selection method based on mutual information and random projection. Finally, only 4k features are kept and used for training a deep feed forward neural network. MtNet has been evaluated on a dataset containing 6.5 M samples (2.8 M malicious and 3.7 M benign) and was able to achieve a 0.36% and 2.94% error rates, in the binary malware detection problem and the malware classification problem, respectively.

In [50] a neural network architecture consisting of CNN and feed-forward neural network (FFNN) for malware families classification is proposed. In this work, they opted for a static analysis of executables, more precisely the PE header metadata, the PE imports, and the opcode sequences. The proposed system has been evaluated on a dataset composed of 22,757 samples (22,694 malicious and 63 benign executables) and was able to achieve an F-score of 92% of along with a precision and recall of 93%.

Kown et al. [53] employed RNN on API call sequences belonging to nine (09) different malware families in order to generate behavioral patterns that can allow to distinguish between these different families. They used Jaccard similarity measure compared with the generated APIs patterns with those of extracted from test samples. They achieved an average classification accuracy of 71%.

With the aim of detecting and categorizing malware into their respective families, Vinaykumar et al. [90] designed a scalable and hybrid approach that combines visualization and deep learning architectures for static, dynamic, and image processing-based. The proposed approach is called ScaleMalNet and executes the task into two phases. In the first one, various machine learning and deep learning algorithms are employed for static and dynamic malware analysis. In addition, detection of malware from images using deep learning is evaluated where the file is converted into an image.

Table 6.7 Summary of malware classification solutions that employ dynamic features

Ref.	*Year*	*Features used*	*Feature representation*	*Dataset size*	*DL Algo.*	*Results*
[22]	2015	API, exec traces, net. traffic	Vector	1800	DBN	Acc: 98.6%
[49]	2016	APIs	1-hot vect, n-gram, sequences	4753	CNN + LSTM	Acc: 89.4%
[41]	2016	APIs + Args NullTerObj	n-grams, sequences	6.5M	FFN	ER: 0.23%, 2.94%
[53]	2017	API calls	Sequence	787	RNN, Similarity	Acc: 71%
[90]	2019	PE Metadata, bytecode	Image	≈300K	DNN, CNN, LSTM	Acc: 98.9%

In the second stage, malware were grouped into corresponding malware families using image-processing approaches. Various experimental tests on both the publicly and privately collected datasets indicated that deep learning-based methodologies outperformed classical machine learning algorithms. In Table 6.7, we provide a summary of the discussed malware classification solutions that employ dynamic features.

6.7 Open Issues and Future Directions

In this section, we identify the following main open challenges and future research directions with respect to malware analysis using deep learning:

- *Concept drift problem and deep learning model update*: Machine and deep learning models assume that training data follow a stationary distribution, and this distribution is valid for new data. However, malware are continuously evolving in order to evade detection, which is known as "concept drift," and defined as changing of relationships in the data, and hence the performance of the learning models decreases over time. Thus, the main challenge is how to define patterns from malware that can resist to malware evolution for a long time. Also, there is a need to continuously adapt the deep learning model through full, partial, or incremental learning to detect new variants of known malware or zero-day malware. In this case, the challenge is how to reduce the cost of continuous learning model update, especially if the updated model needs to be transferred to machines at large-scale.

- *Manual malware investigation*: The high number of false positives generated by deep learning malware detection systems can be a major issue, as it imposes on security analysts spending time on investigating false alarms.
- *Comparison of anti-malware solutions using deep learning*: In the literature, the deep learning models for malware analysis are tested under different datasets and different validation and experimental settings. Thus, it is not possible to provide a fair comparison among the state-of-the-art solutions. To deal with this issue, researchers should share their datasets or conduct experiments on common datasets, such as Kaggle Microsoft Malware Prediction [63].
- *Explainable deep learning model*: The deep learning models are considered as black boxes and produce unexplainable predictions, which limit their acceptability and their adoption in anti-malware products. In order to make these models interpretable, few works that extract rules from deep neural networks are proposed in the context of malware analysis, which extract the embedded knowledge in the DNN in the form of explainable rules [59, 96]. Further efforts could be made to make to focus on explainable anti-malware deep learning models.
- *Adversarial learning*: Malware developers could identify ways to evade detection. To this end, they deceive the deep learning model by injecting adversarial inputs, i.e., samples that are subject to feature perturbations, which induce feature representations that are close to benign samples and cause the deep learning model to make wrong decisions. Future research should focus on fortifying and testing the deep learning models against adversarial samples through different techniques, such as data augmentation from Generative Adversarial Networks (GANs).
- *Collaborative anti-malware solutions using deep learning*: Deep learning models require huge amount of data to train the model, and they are generated in high-computing infrastructures like a cloud or a server. Hence, there is a need for the server and the different client machines to work in a collaborative manner through different forms, such as sharing of pre-trained global model and sharing of local model's parameter between the client and the server. The above collaborative forms raise privacy concerns like exposing the local data

and models. To deal with this issue, it is recommended that future research efforts on malware analysis using deep learning consider some collaborative schemes, such as deep learning model splitting and deferentially private model parameters [13].

6.8 Conclusion

In this paper, we surveyed the state-of-the-art solutions for Windows malware analysis using deep learning. We first provided the necessary background information regarding malware analysis as well as deep learning. We then introduced our proposed taxonomy, and we discussed the existing solutions with regard to this taxonomy.

In conclusion, we believe that deep learning can be extremely effective in malware analysis and detection, especially when dealing with obfuscated and zero-day malware. However, it is important to remember that no single solution is perfect, and there are always trade-offs to be made. For example, deep learning models may require more resources to train and deploy than conventional machine learning solutions or signature-based approaches. Additionally, deep learning models may be more susceptible to adversarial attacks. Therefore, it is important to carefully consider the risks and benefits of deploying a deep learning model for malware analysis.

References

1. Aditya WR, Hadiprakoso RB, Waluyo A et al (2021) Deep learning for malware classification platform using windows API call sequence. In: 2021 international conference on informatics, multimedia, cyber and information system (ICIMCIS). IEEE, Piscataway, pp 25–29
2. Alzubaidi L, Zhang J, Humaidi AJ, Al-dujaili A, Duan Y, Al-Shamma O, Santamaría J, Fadhel MA, Al-Amidie M, Farhan L (2021) Review of deep learning: concepts, CNN architectures, challenges, applications, future directions. J Big Data 8:1–74
3. Andrade EDO, Viterbo J, Vasconcelos CN, Guérin J, Bernardini FC (2019) A model based on LSTM neural networks to identify five different types of malware. Proc Comput Sci 159:182–191
4. Aslan Ö, Yilmaz AA (2021) A new malware classification framework based on deep learning algorithms. IEEE Access 9:87936–87951

5. Aslan ÖA, Samet R (2020) A comprehensive review on malware detection approaches. IEEE Access 8:6249–6271
6. Athiwaratkun B, Stokes JW (2017) Malware classification with LSTM and GRU language models and a character-level CNN. In: 2017 IEEE international conference on acoustics, speech and signal processing (ICASSP). IEEE, Piscataway, pp 2482–2486
7. Belaoued M, Mazouzi S (2015) A real-time pe-malware detection system based on chi-square test and pe-file features. In: IFIP international conference on computer science and its applications. Springer, Berlin, pp 416–425
8. Belaoued M, Mazouzi S (2016) A chi-square-based decision for real-time malware detection using pe-file features. J Inform Process Syst 12(4):644–660
9. Belaoued M, Boukellal A, Koalal MA, Derhab A, Mazouzi S, Khan FA (2019) Combined dynamic multi-feature and rule-based behavior for accurate malware detection. Int J Distrib Sensor Netw 15(11):1550147719889907
10. Belaoued M, Derhab A, Mazouzi S, Khan FA (2020) Macomal: a multi-agent based collaborative mechanism for anti-malware assistance. IEEE Access 8:14329–14343
11. Berman DS, Buczak AL, Chavis JS, Corbett CL (2019) A survey of deep learning methods for cyber security. Information 10(4):122
12. Bougueroua N, Mazouzi S, Belaoued M, Seddari N, Derhab A, Bouras A (2021) A survey on multi-agent based collaborative intrusion detection systems. J Artif Intell Soft Comput Res 11(2):111–142
13. Boulemtafes A, Derhab A, Challal Y (2020) A review of privacy-preserving techniques for deep learning. Neurocomputing 384:21–45
14. Breiman L (2001) Random forests. Mach Learn 45(1):5–32
15. Cao L, ZhiMin (2019) An overview of deep reinforcement learning. In: Proceedings of the 2019 4th international conference on automation, control and robotics engineering
16. Caron M, Bojanowski P, Joulin A, Douze M (2018) Deep clustering for unsupervised learning of visual features. CoRR abs/1807.05520

17. Cavnar WB, Trenkle JM et al (1994) N-gram-based text categorization. Ann Arbor MI 48113(2):161–175
18. Choi S, Jang S, Kim Y, Kim J (2017) Malware detection using malware image and deep learning. In: 2017 international conference on information and communication technology convergence (ICTC), IEEE, Piscataway, pp 1193–1195
19. Cui Z, Du L, Wang P, Cai X, Zhang W (2019) Malicious code detection based on CNNs and multi-objective algorithm. J Parallel Distrib Comput 129:50–58
20. Darabian H, Homayounoot S, Dehghantanha A, Hashemi S, Karimipour H, Parizi RM, Choo KKR (2020) Detecting cryptomining malware: a deep learning approach for static and dynamic analysis. J Grid Comput 18(2):293–303
21. Darem A, Abawajy J, Makkar A, Alhashmi A, Alanazi S (2021) Visualization and deep-learning-based malware variant detection using opcode-level features. Fut Gener Comput Syst 125:314–323
22. David OE, Netanyahu NS (2015) Deepsign: deep learning for automatic malware signature generation and classification. In: 2015 international joint conference on neural networks (IJCNN). IEEE, Piscataway, pp 1–8
23. Davuluru VSP, Narayanan BN, Balster EJ (2019) Convolutional neural networks as classification tools and feature extractors for distinguishing malware programs. In: 2019 IEEE national aerospace and electronics conference (NAECON). IEEE, Piscataway, pp 273–278
24. Deng L, Yu D et al (2014) Deep learning: methods and applications. Found Trends® Signal Process 7(3–4):197–387
25. Ding Y, Chen S, Xu J (2016) Application of deep belief networks for opcode based malware detection. In: 2016 international joint conference on neural networks (IJCNN). IEEE, Piscataway, pp 3901–3908
26. Egele M, Scholte T, Kirda E, Kruegel C (2012) A survey on automated dynamic malware-analysis techniques and tools. ACM Comput Surv 44(2):6
27. Eilam E (2011) Reversing: secrets of reverse engineering. Wiley, New York

28. Eskandari M, Hashemi S (2011) Metamorphic malware detection using control flow graph mining. Int J Comput Sci Netw Secur 11(12):1–6
29. Geurts P, Ernst D, Wehenkel L (2006) Extremely randomized trees. Mach Learn 63(1):3–42
30. Ghanei H, Manavi F, Hamzeh A (2021) A novel method for malware detection based on hardware events using deep neural networks. J Comput Virol Hacking Tech 17(4):319–331
31. Gibert D, Mateu C, Planes J (2020) Hydra: a multimodal deep learning framework for malware classification. Comput Secur 95:101873
32. Gibert D, Mateu C, Planes J (2020) The rise of machine learning for detection and classification of malware: research developments, trends and challenges. J Netw Comput Appl 153:102526
33. Guarnieri C, Tanasi A, Bremer J, Schloesser M (2012) The cuckoo sandbox
34. Hahn K (2014) Robust static analysis of portable executable malware. HTWK Leipzig
35. Hailat Z, Komarichev A, Chen XW (2018) Deep semi-supervised learning. In: 2018 24th international conference on pattern recognition (ICPR), pp 2154–2159
36. Hardy W, Chen L, Hou S, Ye Y, Li X (2016) DL4MD: a deep learning framework for intelligent malware detection. In: Proceedings of the international conference on data mining (DMIN), The Steering Committee of The World Congress in Computer Science, Computer Engineering and Applied Computing (WorldComp), p 61
37. He K, Zhang X, Ren S, Sun J (2015) Delving deep into rectifiers: surpassing human-level performance on imagenet classification. In: Proceedings of the IEEE international conference on computer vision, pp 1026–1034
38. Hemalatha J, Roseline SA, Geetha S, Kadry S, Damaševičius R (2021) An efficient densenet-based deep learning model for malware detection. Entropy 23(3):344
39. Hex-Rays S (2008) Ida pro disassembler

40. Hua Y, Du Y, He D (2020) Classifying packed malware represented as control flow graphs using deep graph convolutional neural network. In: 2020 international conference on computer engineering and application (ICCEA). IEEE, Piscataway, pp 254–258
41. Huang W, Stokes JW (2016) Mtnet: a multi-task neural network for dynamic malware classification. In: Detection of intrusions and malware, and vulnerability assessment. Springer, Berlin, pp 399–418
42. Jain M, Andreopoulos W, Stamp M (2020) Convolutional neural networks and extreme learning machines for malware classification. J Comput Virol Hacking Tech 16(3):229–244
43. Jian Y, Kuang H, Ren C, Ma Z, Wang H (2021) A novel framework for image-based malware detection with a deep neural network. Comput Secur 109:102400
44. Kalash M, Rochan M, Mohammed N, Bruce ND, Wang Y, Iqbal F (2018) Malware classification with deep convolutional neural networks. In: 2018 9th IFIP international conference on new technologies, mobility and security (NTMS). IEEE, Piscataway, pp 1–5
45. Kambar MEZN, Esmaeilzadeh A, Kim Y, Taghva K (2022) A survey on mobile malware detection methods using machine learning. In: 2022 IEEE 12th annual computing and communication workshop and conference (CCWC). IEEE, Piscataway, pp 0215–0221
46. Kan Z, Wang H, Xu G, Guo Y, Chen X (2018) Towards light-weight deep learning based malware detection. In: 2018 IEEE 42nd annual computer software and applications conference (COMPSAC), vol 1. IEEE, Piscataway, pp 600–609
47. Kang J, Jang S, Li S, Jeong YS, Sung Y (2019) Long short-term memory-based malware classification method for information security. Comput Electr Eng 77:366–375
48. Kebede TM, Djaneye-Boundjou O, Narayanan BN, Ralescu A, Kapp D (2017) Classification of malware programs using autoencoders based deep learning architecture and its application to the microsoft malware classification challenge (big 2015) dataset. In: 2017 IEEE national aerospace and electronics conference (NAECON). IEEE, Piscataway, pp 70–75

49. Kolosnjaji B, Zarras A, Webster G, Eckert C (2016) Deep learning for classification of malware system call sequences. In: Australasian joint conference on artificial intelligence. Springer, Berlin, pp 137–149
50. Kolosnjaji B, Eraisha G, Webster G, Zarras A, Eckert C (2017) Empowering convolutional networks for malware classification and analysis. In: 2017 international joint conference on neural networks (IJCNN). IEEE, Piscataway, pp 3838–3845
51. Kouliaridis V, Kambourakis G (2021) A comprehensive survey on machine learning techniques for android malware detection. Information 12(5):185
52. Kumari M, Hsieh G, Okonkwo CA (2017) Deep learning approach to malware multi-class classification using image processing techniques. In: 2017 international conference on computational science and computational intelligence (CSCI). IEEE, Piscataway, pp 13–18
53. Kwon I, Im EG (2017) Extracting the representative API call patterns of malware families using recurrent neural network. In: Proceedings of the international conference on research in adaptive and convergent systems, pp 202–207
54. Le Q, Boydell O, Mac Namee B, Scanlon M (2018) Deep learning at the shallow end: malware classification for non- domain experts. Digit Invest 26:S118–S126
55. LeCun Y, Bengio Y, Hinton G (2015) Deep learning. Nature 521(7553):436
56. Li C, Lv Q, Li N, Wang Y, Sun D, Qiao Y (2022) A novel deep framework for dynamic malware detection based on API sequence intrinsic features. Comput Secur 116:102686
57. Lin WC, Yeh YR (2022) Efficient malware classification by binary sequences with one-dimensional convolutional neural networks. Mathematics 10(4):608
58. Liu K, Xu S, Xu G, Zhang M, Sun D, Liu H (2020) A review of android malware detection approaches based on machine learning. IEEE Access 8:124579–124607
59. Mahdavifar S, Ghorbani AA (2020) Dennes: deep embedded neural network expert system for detecting cyber attacks. Neural Comput Appl 32(18):14753–14780

60. Maniath S, Ashok A, Poornachandran P, Sujadevi V, AU PS, Jan S (2017) Deep learning LSTM based ransomware detection. In: 2017 recent developments in control, automation & power engineering (RDCAPE). IEEE, Piscataway, pp 442–446
61. Masabo E, Kaawaase KS, Sansa-Otim J (2018) Big data: deep learning for detecting malware. In: 2018 IEEE/ACM symposium on software engineering in Africa (SEiA). IEEE, Piscataway, pp 20–26
62. Meng X, Shan Z, Liu F, Zhao B, Han J, Wang H, Wang J (2017) Mcsmgs: malware classification model based on deep learning. In: 2017 international conference on cyber-enabled distributed computing and knowledge discovery (CyberC), pp 272–275
63. Microsoft malware prediction (2018). https://www.kaggle.com/c/microsoft-malware-prediction
64. Mohammed TM, Nataraj L, Chikkagoudar S, Chandrasekaran S, Manjunath B (2021) Malware detection using frequency domain-based image visualization and deep learning. arXiv preprint arXiv:210110578
65. Mourtaji Y, Bouhorma M, Alghazzawi D (2019) Intelligent framework for malware detection with convolutional neural network. In: Proceedings of the 2nd international conference on networking, information systems & security, pp 1–6
66. Narayanan BN, Davuluru VSP (2020) Ensemble malware classification system using deep neural networks. Electronics 9(5):721
67. Ni S, Qian Q, Zhang R (2018) Malware identification using visualization images and deep learning. Comput Secur 77:871–885
68. Pan Y, Ge X, Fang C, Fan Y (2020) A systematic literature review of android malware detection using static analysis. IEEE Access 8:116363–116379
69. Pascanu R, Stokes JW, Sanossian H, Marinescu M, Thomas A (2015) Malware classification with recurrent networks. In: 2015 IEEE international conference on acoustics, speech and signal processing (ICASSP). IEEE, Piscataway, pp 1916–1920
70. Patterson J, Gibson A (2017) Deep Learning: a practitioner's approach. O'Reilly Media

71. Pietrek M (1994) Peering inside the pe: a tour of the win32 (r) portable executable file format. Microsoft Systems Journal-US Edition, pp 15–38
72. Qiu J, Zhang J, Luo W, Pan L, Nepal S, Xiang Y (2020) A survey of android malware detection with deep neural models. ACM Comput Surv 53(6):1–36
73. Raff E, Sylvester J, Nicholas C (2017) Learning the PE header, malware detection with minimal domain knowledge. arXiv preprint arXiv:170901471
74. Rahul R, Anjali T, Menon VK, Soman K (2017) Deep learning for network flow analysis and malware classification. In: International symposium on security in computing and communication. Springer, Berlin, pp 226–235
75. Sahin M, Bahtiyar S (2020) A survey on malware detection with deep learning. In: 13th international conference on security of information and networks, pp 1–6
76. Saxe J, Berlin K (2015) Deep neural network based malware detection using two dimensional binary program features. In: 2015 10th international conference on malicious and unwanted software (MALWARE). IEEE, Piscataway, pp 11–20
77. Schultz MG, Eskin E, Zadok E, Stolfo SJ (2001) Data mining methods for detection of new malicious executables. In: Security and privacy, 2001. S&P 2001. Proceedings. 2001 IEEE Symposium on. IEEE, Piscataway, pp 38–49
78. Shafiq MZ, Tabish SM, Mirza F, Farooq M (2009) Pe-miner: mining structural information to detect malicious executables in realtime. In: International workshop on recent advances in intrusion detection. Springer, Berlin, pp 121–141
79. Shibahara T, Yagi T, Akiyama M, Chiba D, Yada T (2016) Efficient dynamic malware analysis based on network behavior using deep learning. In: 2016 IEEE global communications conference (GLOBECOM). IEEE, Piscataway, pp 1–7
80. Siddiqui M, Wang MC, Lee J (2008) A survey of data mining techniques for malware detection using file features. In: Proceedings of the 46th annual southeast regional conference on xx. ACM, New York, pp 509–510

81. Singh J, Singh J (2021) A survey on machine learning-based malware detection in executable files. J Syst Archit 112:101861
82. Souri A, Hosseini R (2018) A state-of-the-art survey of malware detection approaches using data mining techniques. Hum.-Centric Comput Inform Sci 8(1):1–22
83. Stevenson M, Mues C, Bravo C (2021) Deep residential representations: using unsupervised learning to unlock elevation data for geo-demographic prediction. CoRR abs/2112.01421
84. Tian D, Ying Q, Jia X, Ma R, Hu C, Liu W (2021) Mdchd: a novel malware detection method in cloud using hardware trace and deep learning. Comput Netw 198:108394
85. Tobiyama S, Yamaguchi Y, Shimada H, Ikuse T, Yagi T (2016) Malware detection with deep neural network using process behavior. In: 2016 IEEE 40th annual computer software and applications conference (COMPSAC), vol 2. IEEE, Piscataway, pp 577–582
86. Ucci D, Aniello L, Baldoni R (2019) Survey of machine learning techniques for malware analysis. Comput Secur 81:123–147
87. Urooj U, Al-rimy BAS, Zainal A, Ghaleb FA, Rassam MA (2021) Ransomware detection using the dynamic analysis and machine learning: a survey and research directions. Appl Sci 12(1):172
88. Vasan D, Alazab M, Wassan S, Safaei B, Zheng Q (2020) Image-based malware classification using ensemble of CNN architectures (IMCEC). Comput Secur 92:101748
89. Venkatraman S, Alazab M, Vinayakumar R (2019) A hybrid deep learning image-based analysis for effective malware detection. J Inform Secur Appl 47:377–389
90. Vinayakumar R, Alazab M, Soman K, Poornachandran P, Venkatraman S (2019) Robust intelligent malware detection using deep learning. IEEE Access 7:46717–46738
91. Wang H, Zhu Z, Tong Z, Yin X, Feng Y, Shi G, Meng D (2021) An effective approach for malware detection and explanation via deep learning analysis. In: 2021 international joint conference on neural networks (IJCNN). IEEE, Piscataway, pp 1–10
92. Wang X, Yiu SM (2016) A multi-task learning model for malware classification with useful file access pattern from API call sequence. arXiv preprint arXiv:161005945

93. Weston J, Ratle F, Mobahi H, Collobert R (2012) Deep learning via semi-supervised embedding. Springer, Berlin, pp 639–655
94. Wu Z, Pan S, Chen F, Long G, Zhang C, Philip SY (2020) A comprehensive survey on graph neural networks. IEEE Trans Neural Netw Learn Syst 32(1):4–24
95. Xiao F, Lin Z, Sun Y, Ma Y (2019) Malware detection based on deep learning of behavior graphs. Math Probl Eng 2019:1–10
96. Yan A, Chen Z, Zhang H, Peng L, Yan Q, Hassan MU, Zhao C, Yang B (2021) Effective detection of mobile malware behavior based on explainable deep neural network. Neurocomputing 453:482–492
97. Yan J, Qi Y, Rao Q (2018) Detecting malware with an ensemble method based on deep neural network. Security and Communication Networks 2018
98. Ye Y, Li T, Adjeroh D, Iyengar SS (2017) A survey on malware detection using data mining techniques. ACM Comput Surv 50(3):41
99. Yu D, Deng L (2016) Automatic speech recognition. Springer, Berlin
100. Yuan B, Wang J, Liu D, Guo W, Wu P, Bao X (2020) Byte-level malware classification based on Markov images and deep learning. Comput Secur 92:101740
101. Yue S (2017) Imbalanced malware images classification: a CNN based approach. arXiv preprint arXiv:170808042
102. Yuxin D, Siyi Z (2019) Malware detection based on deep learning algorithm. Neural Comput Appl 31(2):461–472
103. Zhang J, Qin Z, Yin H, Ou L, Hu Y (2016) Irmd: malware variant detection using opcode image recognition. In: 2016 IEEE 22nd international conference on parallel and distributed systems (ICPADS). IEEE, Piscataway, pp 1175–1180

CHAPTER 7

Malware Analysis for IoT and Smart AI-Based Applications

Syed Emad ud Din Arshad, Moustafa M. Nasralla, Sohaib Bin Altaf Khattak, Taqwa Ahmed Alhaj, and Ikram ur Rehman

7.1 Introduction

Recent years have seen a sharp rise in the usage of Internet of Things (IoT) devices in a number of industries, including industry, health, automation,

S. E. ud Din Arshad
National University of Sciences and Technology, Islamabad, Pakistan
e-mail: 14mseesarshad@seecs.edu.pk

M. M. Nasralla (✉) • S. B. A. Khattak
Smart Systems Engineering Lab, Department of Communications and Networks Engineering, Prince Sultan University, Riyadh, Saudi Arabia
e-mail: skhattak@psu.edu.sa; mnasralla@psu.edu.sa

T. A. Alhaj
Information Assurance and Security Research Group, Faculty of Computing, Universiti Teknologi Malaysia, UTM, Johor Bahru, Malaysia

I. ur Rehman
School of Computing and Engineering, University of West London, London, UK
e-mail: ikram.rehman@uwl.ac.uk

I. Almomani et al. (eds.), *Cyber Malware*, Security Informatics and Law Enforcement, https://doi.org/10.1007/978-3-031-34969-0_7

and education, as well as smart homes and smart cities [1, 2]. According to current estimates, there will be 75.44 billion connected IoT devices worldwide by 2025 [3]. According to another study, the next significant step in achieving the Internet's goal of linking the entire world is the network of connected "smart" products [3]. The IoT technology is closely related to our daily life and is applied in several real-life related applications. It has evolved rapidly and now has covered almost every aspect of modern life, with its applications ranging from home-based services to emergency management services and from societal and environmental applications to industrial and technological applications [4]. Under the umbrella of each of these domains lie thousands of use cases and applications, for example, smart living rooms, smart kitchens, smart garages, smart doors, smart cooling, and refrigerating systems, healthcare applications for older people, or any other monitoring, tracking, or reporting systems [2, 5]. Intelligent transportation and traffic management is another example of IoT, which has a significant effect on our lives. The societal applications improve the lifestyle of the general public and bring a lot of services at the tip of their fingers. Security and surveillance have been revolutionized with the advent of IoT, like intrusion detection systems, and smart surveillance systems. Wildlife monitoring, environmental monitoring, smart farming, observing energy consumption patterns, electricity management, water distribution, waste management, smart marketing, and many similar applications are an essential part of our society now. IoT devices are usually connected through the wireless channel because of their flexibility and mobility. Several wireless communication technologies are used for IoT deployment, depending upon the application requirements [4]. These communication technologies can also be classified as long and short range. The most commonly used short-range communication technologies are RFID, Wi-Fi, ZigBee, and Bluetooth. The widely used long-range communication technologies for IoT are Sigfox, LoRaWAN, Weightless, Narrow Band IoT, and Enhanced Machine Type Communication (eMTC) (Fig. 7.1).

The common features among most applications are low cost, low processing, low power, low storage, and low bit rate. Computers, smartphones, communication interfaces, RFID tags, actuators, readers, cameras, controllers, GPS, sensors, operating systems, lightweight services, and preloaded apps generally make the IoT infrastructure. This technology is not as secure as it seems, and it also raises additional security and privacy issues. IoT networks have weak or no security since they rely on inexpensive devices (such temperature sensors, security cameras, etc.) with constrained

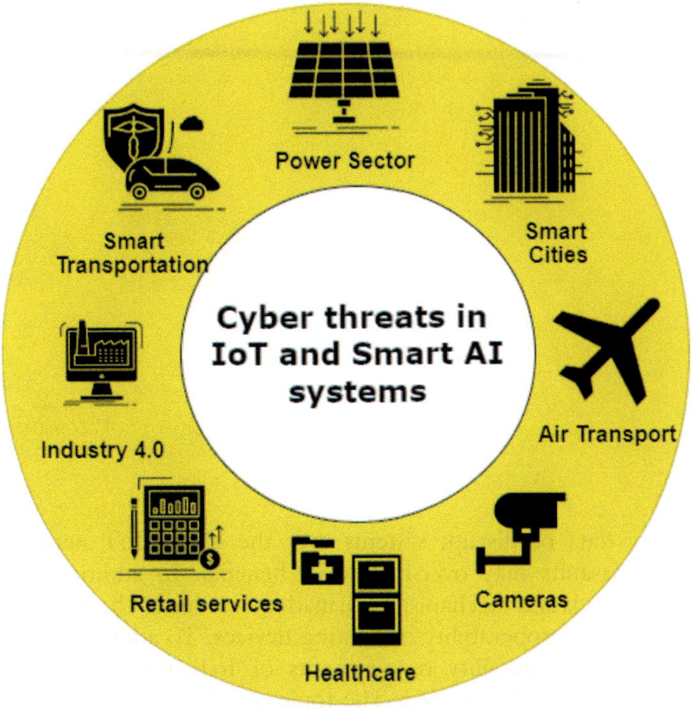

Fig. 7.1 Cyberattacks in IoT

resources (such as low-power sources, limited memory, and computing power) [6]. Due to these restrictions, it is challenging to perform complex security tasks on those devices, making it simple for malicious actors to compromise them and engage in a number of illegal activities that jeopardize the security and integrity of the devices and network [7, 8]. As IoT and artificial intelligence (AI) are enabling a wide range of services across several sectors [9, 10], it makes all these sectors also vulnerable to cyberattacks as shown in Fig. 7.1. Malware encompasses all types of malicious software, such as data theft, snooping, and so on as shown in Fig. 7.2. According to Kaspersky Lab, the virus is a "computer software designed to infect and increase harm to a genuine user's PC" [11].

If the cybersecurity issues are not effectively controlled, hackers will exploit the flaws and vulnerabilities of devices or objects and subsequently

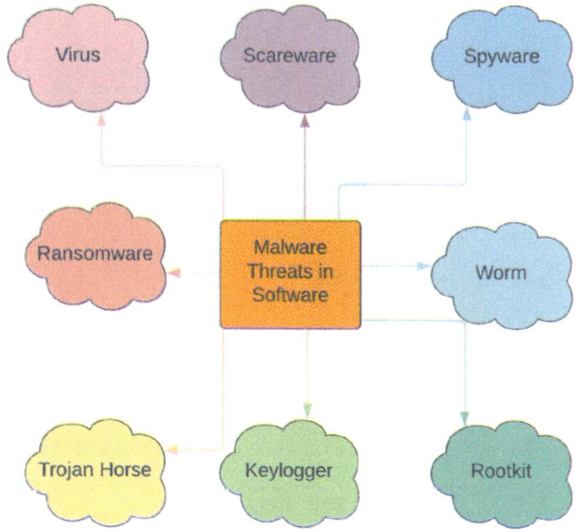

Fig. 7.2 Cyberattacks in IoT

manipulate data or disrupt systems over the global IoT network. IoT faults and assaults may overshadow its benefits. In addition, standard security methods and mechanisms are inadequate due to the low scalability, integrity, and interoperability of existing devices. To address the security, privacy, and dependability requirements of IoT, new approaches and technologies should be created. The topic of cybersecurity challenges on IoT platforms and AI-based applications is a big global concern that necessitates a comprehensive evaluation from both the research and industrial groups. This chapter evaluates security issues that are expected to limit IoT deployment and intends to explore different methods for the detection and evasion of cybersecurity threats in IoT domain. The chapter is structured as follows: Sect. 7.2 discusses the work related to IoT cybersecurity. In Sect. 7.3, the potential threat challenges in relation to IoT applications and services are assessed. In Sect. 7.4 malware attacks and threats are discussed. In Sect. 7.5 malware detection and evasion approaches are presented, and the final section concludes the chapter.

7.2 Related Works

A large number of researchers have discussed the network vulnerabilities and their potential solutions for cyber physical systems. This section covers some of these researches and provide a brief overview of the existing work in this domain.

The authors of [12] examined home automation systems, such as magnetic sensors, motion sensors, and industrial IoT devices, and discovered that smart meters are susceptible to a number of assaults because of the inadequate security measures used during their development and deployment. These gadgets advocate using an encrypted channel as a security mechanism while communicating via an RF (radio frequency) channel. Mutual authentication, the physically unclonable function, finite resources, side-channel analysis, and cloning attacks were all topics covered in the research in [13]. In their article, security mechanisms for protocol verification, session key formation, and mutual authentication are given. Mutual authentication offers useful information about key distribution between devices and during sessions. The suggested remedy lessens the danger of replay and man-in-the-middle attacks. The author of [14] discussed several security difficulties and threats, privacy worries, IoT device integration with the blockchain, and various security research fields. One of the most critical problems in Android/iOS applications is application repackaging. The authors of [15] presented their research on repackaged software. They addressed five issues: (1) the current unfavorable repackaging practices, (2) the way adware is embedded in the code, (3) the kinds of apps used to repackage, (4) the motivations behind people downloading repackaged software, and (5) the way an app's characteristics change in the repackaged version. The drawback of static malware detection techniques, such as TinyDroid, DroidFDR [16], DroidEnsemble, and NsDroid, is that they are not appropriate for dynamic analysis. A quick and efficient Android malware detection tool is NsDroid. NsDroid is 20 times faster than previous graph-based techniques [17, 18] since it is built on a local function graph [19]. The author of [20] cited a dispute between various sensors managed by a smartphone and offered a LOD-based solution (Linked Open Data). LOD makes it possible to more effectively utilize the services and features of the resident's profile while also defining the connections between the various services and items in the home. The authors of the article [21] provided useful insight into the use of signature- and anomaly-based methods for detecting mobile malware.

Authors in [22] discuss the cybersecurity threats in Mobile Adhoc Networks (MANETS), which plays a key function in many IoT settings. MANETS are vulnerable to numerous packet-drop attacks, including as gray- and black-hole attacks. The authors looked at numerous blackhole attack types and employed learning, cooperative, and other detection strategies. Their study concludes that trust-based scheme performs better when compared to other schemes. For availability, security, and reliability of MANETS, the threat of botnets must be taken care of. The newly developed botnets are designed to dodge the detection systems. Large amounts of data processing are required for high computational requirements to differentiate between normal and botnet traffic. The authors in [23] proposed a system to address this problem by developing a scalable and decentralized framework, based on characterization of the behavior of legitimate hosts, and detect unseen botnet traffic.

Cross-architecture detection of IoT malware is a very challenging task because these IoT devices are very heterogeneous. A solution to this problem is proposed by using graph-based malware detection methods to detect malware in IoT devices [24]. Graph-based techniques detected complicated and zero-day malicious codes with greater accuracy. MalInsight, [25] a malware detection system, breaks down malware into three categories: basic structure, low-level behavior, and high-level behavior. Operations were carried out based on the three elements on files, structural features, networks, and registries. The framework might quickly identify malware that hasn't been seen and make it simple for future researchers to find spyware.

Wang et al. [26] utilized lightweight network analysis and machine learning to develop a framework for malware identification in Android devices. In this work, authors combine machine learning with network traffic analysis on the server-side, with minimum resource consumption and minimum impact on the user experience. For the purpose of identifying cyber vulnerabilities and threats, a unique machine learning-based methodology was put forth by [27] to identify cyber threats using novel machine learning-based framework. This framework used observed attack patterns, and in result it was able to identify and detect cyberattacks. Another machine learning technique based on hamming distance is used for malware detection [28]. This method made use of k-medoid-based nearest neighbors (KMNN), weighted all nearest neighbors (WANN), and first nearest neighbors (FNN). These algorithms, which have high recall and precision rates, were employed to identify malicious software. A classification model is proposed to detect mobile malware attacks in IoT systems [29]. Mobile malware attacks are mainly caused because of fraudulent mobile applications and injected malicious applications. Other machine learning techniques for malware detection adopted in IoT and

AI-based smart systems are decision tree, SVM, random forest, and logistic regression [30, 31].

Other popular techniques for malware detection are sandbox environment techniques, blockchain technology, and deep learning. Sandbox is a testing environment used to investigate malware behaviors [32]. Kachare et al. [33] propose a concept for a sandbox environment that analyzes malware, produces reports automatically, and fixes issues. In order to study malware at three different levels–static malware analysis, real-time malware analysis, and network analysis–the suggested model employs multiple machine learning algorithms. Advanced persistent threats (APTs) are immune to anti-malware and anti-virus systems along with other conventional security systems. Advanced evasive techniques are used to tackle these malwares. The work in [34] measures the divergence from a program's typical behavior utilizing Analysis Evasion Malware Sandbox to discover malware evasive behavior (AEMS). Blockchain uses its principles of cryptography, decentralization, and consensus for security. In [35], a blockchain-based malware detection technique leveraging shared signatures of suspicious malware files is put out. With the help of this technique, users can quickly respond to the growing threat of malware by sharing the signatures of dubious files. Deep learning is a part of machine learning family and has been widely used recently in wide range of applications including cybersecurity [36]. Authors in [37] develop a tool to detect IoT-malware infections in smart home networks. It analyzes IoT traffic as captured by means of a spoofing technique.

7.3 Cybersecurity Threats for IoT and Smart AI Applications

The rise of smart cities is made possible by the hyper-connectivity and constant availability of IoT technologies, but they also raise cybersecurity threats and attacks [38]. Approximately 300 percent more cyberattacks on IoT devices were reported in 2019 according to a Forbes review of security incidents [39]. The following are some examples of cyberattacks that could occur in a smart city:

- Traffic light control: because wireless networks have made traffic signals more susceptible to attack, attackers are now able to modify traffic lights and cause accidents [40].

- Attacks on smart vehicles: Attackers can simulate other vehicles in the environment or insert bogus routes into smart vehicles to induce crashes [41].
- Power grid collapse: Attackers might knock out the city's electricity through a power grid collapse [42].
- Water supply:Attackers may alter the concentrations of chemical additives in the water, endangering the public's health [43].
- Surveillance cameras: Attackers can eavesdrop on people and steal personal information via surveillance cameras [44].

Furthermore, based on Hassija [45], the development of IoT solutions raised concerns about data privacy.Data consent is linked to data privacy. Devices or sensors, such cameras on the street or motion sensors under a patient's bed, may capture people's data without their permission. In other words, it's possible that people aren't aware that their data is being gathered. They might also have a vague idea of what information is gathered, how it is kept and processed, and who gains from it. The ambiguity in data gathering and use may jeopardize people's privacy and confidence [46]. The problem of data ownership and the benefits gained through the chain of IoT applications is also related to the data privacy issue [47].

For several years, cities have been the target of security attacks all over the world. For example, in 2015, cyberattacks caused a power outage in Kyiv (a Ukrainian city), depriving residents of electricity for 1 hour [39]. In 2019, ransomware infected the computers of the city administration in Baltimore, USA, and demanded 13 bitcoins in compensation for the decryption of files [48]. Cyberattacks have a negative effect on the technical axis, the city's economy, the quality of life, and more. When cities lose control of their systems, people's lives could also be in danger.

7.4 Malware Attacks in IoT

Software that assists malicious attackers in achieving their objectives is known as malware [49]. It was developed to aid attackers in achieving their goals. These goals include interfering with system operations, gaining access to computer system and network resources, and gathering private information about users without their consent [50]. As a result, malware regularly puts users' privacy, the integrity of the hosts, and the availability of the Internet at danger. Based on the program's execution characteristics, malware is categorized. Additionally, malware is categorized according to its payload, how it exposes or exploits the system, and how it spreads. As a result, malware can be classified into different types, as shown in Fig. 7.2 and discussed below.

7.4.1 Malware Threats in Software

The most dangerous threat to IoT systems is malware since it can either damage the device or, in some situations, change its state to one where the attacker has full control [51]. According to cyberattack statistics, the most well-known malware in IoT software are [52]:

- A rootkit: One of the most deadly kinds of malicious software is a rootkit. With the aid of a rootkit, hackers can get remote access to (and control over) a computer or IoT device without being noticed by the user or security devices. Without the user or security devices realizing it, hackers can use rootkit to remotely access (manage) a machine (i.e., an IoT device). Once installed, the hacker has the ability to remotely execute files, steal important information, modify system settings, and alter the functionality of security software [53].Its stealthiness makes it incredibly challenging to identify and prevent. A rootkit will always attempt to conceal its existence, making security tools incapable of finding and eliminating it. Due to this, it is detected manually using techniques including static analysis, signature scanning, and machine behavior (behavior-based detection) [53].
- Viruses: This harmful software can reproduce itself and infect other systems. By affixing itself to different programs, it spreads to other computers by running the code when a user launches one of the infected apps [53].Because they need their "host" programs to be activated in order to function, viruses cannot run independently. The experimental self-replicating programs known as Bob Thoma's Creeper virus was first identified in the early 1970s [50].
- Worm:An autonomous computer software is called a "worm." A worm may transfer a completely functional clone of itself to other devices, which is important to know. The Morris worm was the first program reported to exhibit worm-like behavior [50].
- Spyware: A form of software known as spyware tracks user behavior without their consent. It integrates itself into a conventional program

by taking advantage of software weaknesses. It might also change the security preferences of the software [53].

- Ransomware: One of the most prevalent types of malware in recent years is ransomware. It's a different kind of virus that controls a gadget (in this case, an IoT device) and demands money from its owner (ransom). By locking the system or encrypting contents on the hard drive, it stops users from accessing the computer. Messages asking the user to pay the ransom to the malware's owner are then displayed by the infection. The key to unlock the hard drive's encrypted files is then given by the ransomware's creator. Typically, it spreads through downloaded files or other network or system software bugs [53].
- Trojan horse: To trick users into downloading and installing it, this spyware impersonates a trustworthy program. It makes it possible for a hacker to get approved remote access to a compromised system. Once a hacker has access to a compromised system, they can take valuable data (e.g., financial information such as account numbers and credit card numbers). To perform even more nefarious actions, it has the ability to install more malicious software on the system.
- Keylogger: A keylogger allows a hacker to monitor a user's keystrokes. Passwords, IDs, and other login details are all logged along with anything else a user puts on the keyboard. A dictionary or brute force assault cannot compete with a key logger attack. By tricking users into downloading the malicious application by clicking on a link in an email, this dangerous program tries to access their device. It is one of the most serious malwares, and you cannot avoid it even with a strong password [53].
- Scareware: Scareware is a new breed of malicious software that tries to convince users to purchase and download pointless and potentially harmful software, including phoney antivirus protection, endangering their financial and personal information [50].

7.4.2 *Malware Threats in Hardware*

Attackers have learned how to operate at the chip level, a crucial part of any system, when it comes to malware in hardware. A device or system can be exposed through a variety of methods. Multiple attacks could be caused by minor modifications to a chip [51].Furthermore, IoT devices are visible in public due to their clear structure. As a result, the system is exposed to multiple threats, including the device ID or serial number of IoT devices being exposed. Meanwhile, the majority of IoT devices are linked to cloud infrastructure, which raises serious security concerns. Therefore, it's very likely that an attacker could compromise the cloud by sending a single piece of malware to multiple IoT devices at the same time [54]. Accordingly, an example of malware-affected hardware is listed below:

- Rowhammer: Rowhammer enables hostile actors to manipulate memory in order to steal information from weak systems, including passwords. The issue has been identified in DDR3 and DDR4 DRAM chips and, when used in conjunction with other attacks, allows users of systems employing the chips to access the data stored in memory.

7.4.3 Malware Threats in Network Communication

Network services are one of the platforms that IoT devices may be attacked over [54]. IoT systems are more susceptible to data leakage attacks as a result of complex encryption algorithms' inability to function on these platforms. A system may be vulnerable to malware infection through distributed denial of service (DDoS) attacks if normal data traffic is not recognized [55]. Additionally, IoT devices are not required to perform payload verification or integrity checks due to resource limitations like computational power and data storage capacity, which encourages IoT device security issues. The most well-known malware in IoT network communication are:

- Bots: A bot is a malicious application that gives its owner remote access to an infected system. Bots are frequently distributed by taking advantage of software bugs and other social engineering techniques. Once a system is compromised, the bot master can utilize Trojans, malware, and worms to turn the affected systems into a botnet. Botnets are frequently used in DDoS assaults, the distribution of spam emails, and phishing scams. Agobot and Sdbot are two of the most well-known bots [50].
- Mirai:Mirai creates a botnet out of networked devices running Linux in order to launch extensive network attacks. Mirai-infected devices are constantly searching the Internet for the common IP addresses of IoT devices. In order to connect into vulnerable IoT devices and infect them with its malware once it has detected them, Mirai employs a collection of common factory default passwords and usernames [56].

- Hajime: Hajime and Mirai are comparable in that both use username and password tables to spread via unsecured open Telnet ports. Unlike Mirai, Hajime is a part of a peer-to-peer network. The controller issues commands to its peer network, and over time, the peer network spreads the message to all other peers. This has a strong design, making it more challenging to knock it over. Aside from design, Hajime has a few other benefits over Mirai. Hajime takes several steps to conceal its operating processes and data on file systems, making it more stealthy [56].

7.5 Malware Detection and Evasion Approaches

The process of evaluating the program's content to determine if the assessed software is malicious or benign is known as malware detection. Three phases are used to detect malware:malware analysis, feature extraction, and classifying malware [57]. In recent years, data mining and machine learning methods have been widely used for malware detection [58]. Data mining is the process of obtaining new and meaningful information from massive datasets or databases and is used for extracting malware characteristics. Machine learning (ML) is a collection of algorithms that effectively predicts application outcomes without being explicitly programmed. The objective of machine learning is to transform the input data into acceptable value intervals via statistical analysis. Machine learning can be used to perform many operations on linked data, including classification, regression, and grouping.

As it is necessary to ensure that only authorized users can access system services, authentication is a key requirement for many layers of a smart system. Particularly in smart cities, IoT devices can authenticate communications from control stations, other nodes, and the network itself. Furthermore, new technologies must be developed to guarantee real-time and trustworthy authentication because the amount of authentication data in smart cities is growing quickly.

7.5.1 *Major Malware Detection Approaches in IoT*

According to the vulnerabilities explained in the previous section, an IoT based smart system must be capable of detecting abnormal events in a real-time manner. Predicting about the upcoming cyber threats is important and is better than recovery after the attack. Therefore it is crucial to develop intelligent malware detection and evasion approaches in order to achieve cybersecurity awareness and automatically predict attacks on smart applications [58]. This section describes several defenses for IoT and smart cities applications against cyber malware attacks. The shared defense strategies will help to advance the security measures for facilitating future commercializing of smart city services. Different malware detection approaches are proposed recently as shown in Fig. 7.3; below we present an overview of these malware detection approaches:

- Signature-based malware detection: Signature is a characteristic of malware that consists of the structure of the malware and uniquely identifies it. This detection method is prevalent in commercial antivirus software and is quick and effective at detecting known malwares, but is ineffective for detecting a novel malware. Using obfuscation techniques, malware belonging to the same family can also readily evade signature-based detection. General functionality of this detection method can be seen in Fig. 7.4.
- Behavior-based malware detection: These approaches analyzes the behavior of the program using monitoring tools and identify whether the program is a malware or not. Even if the program codes are

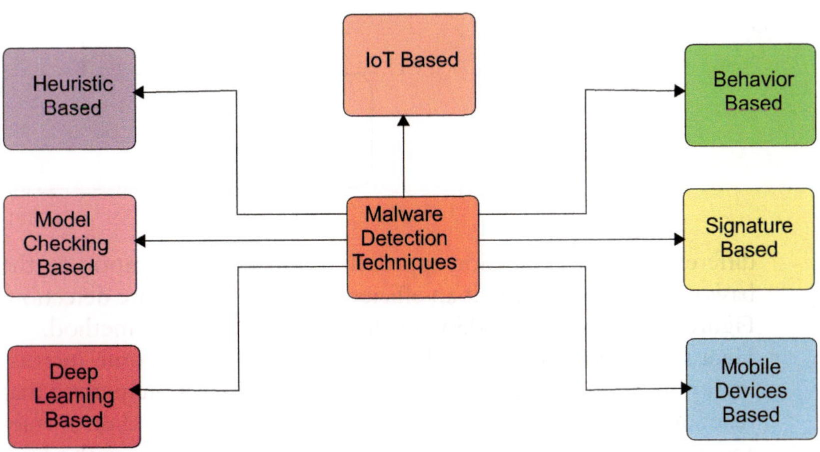

Fig. 7.3 Major malware detection approaches in IoT

Fig. 7.4 Signature-based malware detection

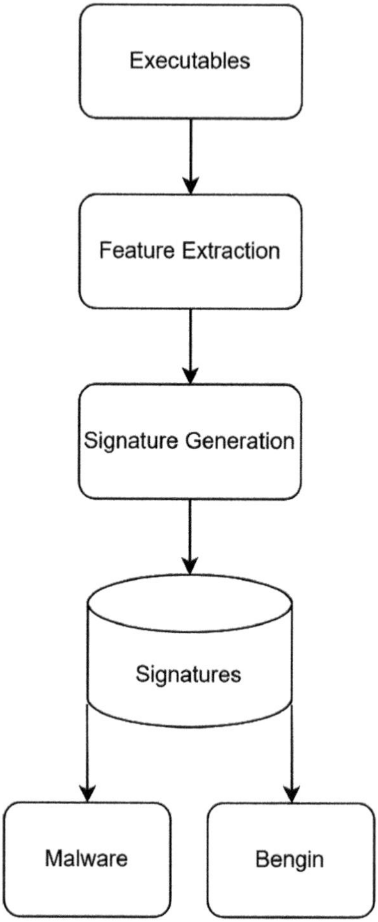

different, still this detection approach can make identification on the basis of behavior making it an efficient method of malware detection. Figure 7.5 shows the working of this malware detection method.

- Heuristic-based detection: These approaches have become increasingly popular in recent years. It's a complicated detection method that relies on past experiences as well as various strategies based on as rules and machine learning [59]. It can detect zero-day malware with a high

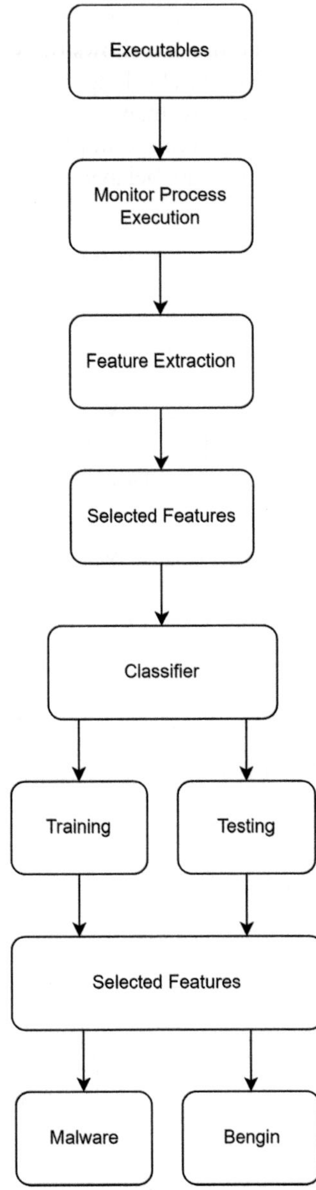

Fig. 7.5
Behavior-based malware detection

degree of accuracy; however, it cannot detect sophisticated malware. Figure 7.6 shows the visual illustration of this method.

- Model checking-based detection: Model checking has been used to identify malware even though its original purpose was to evaluate a system's compliance with standards. In this detection method, malware behaviors are manually extracted, and behavior groups are

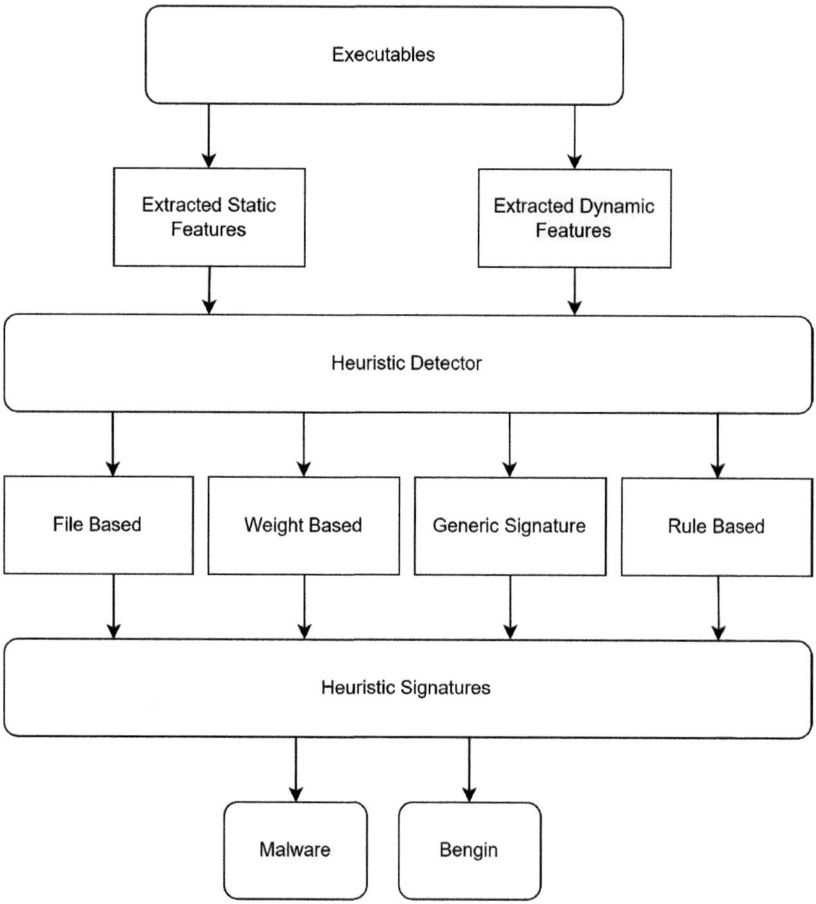

Fig. 7.6 Heuristic-based detection

linear temporal logic (LTL) coded to exhibit a particular trait. CTL and CTPL are two other typical formulas. By examining the flow relationship between one or more system calls and using characteristics like hiding, spreading, and injecting to identify them, program behaviors can be identified. By contrasting these actions, it is possible to tell if the application is malicious or benign. Some new malware can be partially detected using this method, but not all new malware generations. This method is illustrated in Fig. 7.7.

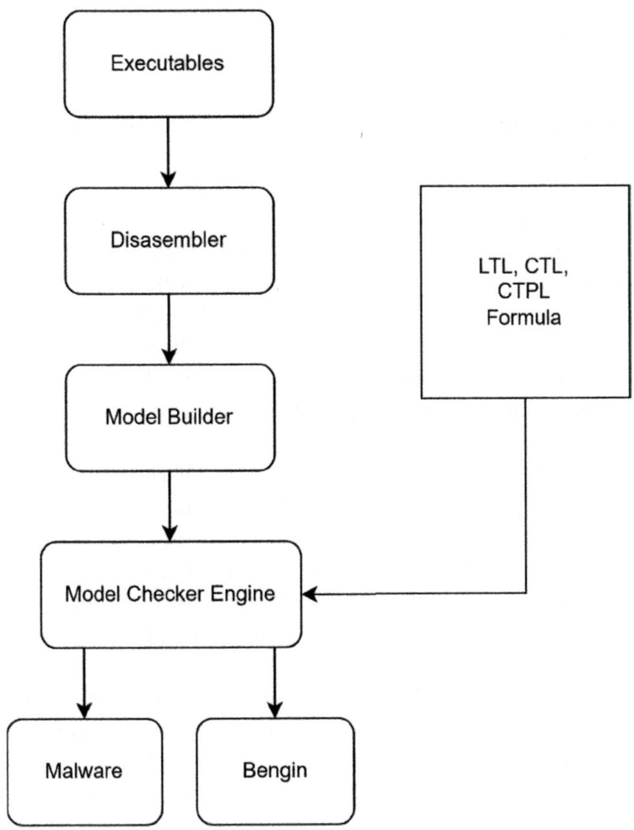

Fig. 7.7 Model checking-based detection

- Deep learning-based detection: Artificial neural networks (ANNs), which learn from examples, are a subset of machine learning (ML) that gave rise to deep learning. Although it is a cutting-edge method that is frequently used for image processing, autonomous driving, and voice control, malware detection rarely makes use of it. It greatly minimizes feature space and is extremely effective, although it is not evasion attack resistant. This approach can be seen in Fig. 7.8.
- Cloud-based detection: Cloud computing is quickly taking off because it has so many advantages, like simple access and on-demand storage. The cloud has been used to detect viruses because of its widespread use. With much larger malware database volumes and intensive computational resources, cloud-based malware detection enhances the performance of desktops and mobile devices in malware detection. Cloud based detection offers security as a service and makes use of a range of detection agents on cloud servers. Any type of file can be submitted, and the user will get a report stating whether or not the file includes malware.
- Mobile device-based detection: In the realm of mobile devices, Android has become the dominant platform. According to recent studies, a new malicious Android app is released every 10 seconds. In light of this, researchers have prioritized Android over other platforms for malware detection. Numerous virus detection approaches, particularly for the Android platform, have been proposed for mobile devices. In general, these technologies identify malware using data mining and machine learning algorithms. Various aspects, including system calls, security-sensitive APIs, information flows, and control flow architectures, are utilized. Even though current research has made progress in detecting old and next-generation malware for mobile devices, detecting complicated malware and scalability of detection algorithms for a big bundle of apps remain formidable challenges.
- IoT-based detection: IoT architecture often consists of a variety of Internet-connected smart devices, such as network cameras, household appliances, and sensors. IoT technology and mobile devices have begun to dominate the Internet more than personal computers. Since mobile and IoT devices are becoming increasingly popular among consumers, they are also becoming increasingly popular among attackers. As a result, the landscape of malware detection schemas is shifting from desktops to IoT and mobile devices.

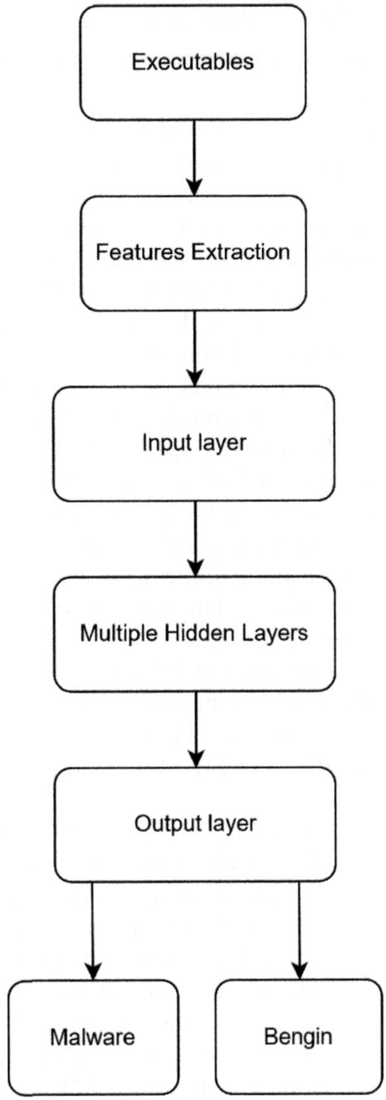

Fig. 7.8 Deep learning-based detection

7.5.2 *Machine Learning Techniques for Malware Detection in IoT*

This section covers learning-based IoT security approaches to detect irregularities in the IoT device activities. In such detection approaches, machine learning (ML) models are trained to identify malicious behavior and then respond accordingly [60]. ML-based approaches outperformed signature-based systems because a minor modification in an attack pattern can readily circumvent a signature-based detection system. However, ML-based systems learn from traffic patterns. They can detect the attack versions with ease [61]. In addition, the CPU load of ML-based systems ranges from low to high because they do not analyze all database signatures. In addition to superior performance in terms of accuracy and speed, ML-based systems capture and reveal the intricate features of attack behavior [62]. These techniques are mainly divided into supervised, unsupervised, and deep learning methods [63]. Supervised ML methods are extensively applied to ensure accuracy and efficiency, whereas unsupervised methods are less common in IoT networks for intrusion detection. Meanwhile, deep learning systems face challenges such as higher computational resource requirements and longer prediction response times. In supervised learning, models are fed with labeled data samples in order to identify the corresponding input-output pattern.

If we have unlabeled data, unsupervised learning techniques are applied. These approaches are predominantly employed for clustering and dimension reduction. Here a quick overview of the aforementioned learning paradigms of ML in the context of IoT security is provided.

7.5.2.1 Brief Description of Commonly Used ML Techniques

- The most extensively used and successful machine learning technology for cybersecurity applications is the support vector machine (SVM). Based on a reference to the margin on either side of the hyperplane, SVM categorizes and separates the two data groups. The margin and separation between hyperplanes can be increased to increase the accuracy of data point detection. SVM needs a lot of memory to operate on data and takes a long time to train. For improved results, SVM should be trained repeatedly to learn the dynamic user's behavior. SVMs were first used in IoT security to distinguish between typical and anomalous behavior. The SVM is used for real-time intrusion detection, with its training pattern being continuously updated, due to its stability, measures, and eligibility.

- Decision tree (DT) is a supervised machine learning technique that utilizes a recursive tree topology. It consists of three compontents: a root or intermediate node, a path, and a leaf node. A tree's root or intermediate node represents an object or property. Each branch of the tree indicates the possible values of the parent node (object). The predicted category or categorized characteristic corresponds to the leaf node. The generated tree is additionally represented as if-then rules. Entropy and information gain metrics are used to choose the ideal intermediate node for further processing as a tree grows. It functions as a classifier either directly or indirectly in IoT intrusion detection techniques.
- K-nearest neighbor (kNN) is an unsupervised learning algorithm. It uses a distance function to calculate the difference/dissimilarity between two data instances. It requires less training time than other classifiers. However, throughout the classification process, its computation time is an overhead. This classifier is based on the concept that data points in the same area that are similar will be closer together than data points that are dissimilar. Based on anomaly scores, there are two primary groups of kNN. The anomaly scores are determined in one of two ways: (1) based on the difference between the kth neighbor and the data point and (2) the density of each data instance that is used to calculate it. The kth data point's value has an impact on the classifier's overall performance.Using KNN, an intrusion detection model for IoT security can be constructed to categorize the normal and abnormal behavior of wireless network sensors in the IoT environment.
- Random forest: An ensemble learning technique called random forest (RF) combines a number of classifiers to produce a problem hypothesis and a typical outcome. It is used to categorize and predict data and is also referred to as a random decision forest. The majority of the time, RF is a collection of predictions from various decision trees. The random forest has been employed in the literature to gauge spam production and identify intrusions. During the model's training phase, it uses less computational power and performs better on nonlinear issues. The decision trees that will be reviewed throughout the prediction process must be picked though, as the random forest combines the predictions of several decision trees.
- The Bayes theorem, which is frequently applied in supervised classification, is the foundation of a Naive Bayes approach. The Bayes theorem uses prior knowledge to describe an incident that might

happen soon. Using previous data, Bayes' technique can identify potential harmful network traffic. In order to show the possibility that a specific characteristic of an unlabeled example matches the labeled feature set under the assumption of feature uniqueness, NB computes the likelihood of specific events using the Bayes' method. NB classifies the properties that may be used during connection protocols and measures the connection status flag in intrusion detection techniques. These traits can be utilized to distinguish between normal and abnormal network behavior.

- Through a series of forward and backpropagation cycles, ANNs are trained. In feedforward, information is supplied into each node of a hidden layer. The activation value of each node in a hidden layer and output layer is established. A classifier's activation function affects how well it performs. Error is calculated using the discrepancy between the network output and the desired value. Using the Guardian Descent method, backpropagation modifies the weights between hidden and output nodes based on this disparity. Up until the necessary level is reached, this process is repeated. Although ANN is easy to use, noise-resistant, and a nonlinear model, it has one drawback in that it requires a lot of training time.
- Forward-looking convolutional neural networks are multi-layer neural networks that are created by extending ANN (CNN). It consists of one or more convolutional layers, one or more fully connected layers, and pooling layers, which are three different types of layers. In order to utilize them at the coarser resolution, it transforms the higher-resolution features into more complex features. CNN is frequently employed in the identification of anomalies and the classification of malicious traffic.

7.5.2.2 Detection and Mitigation Using ML

In order to protect IoT smart furniture from perception-layer assaults, Nasralla et.al [7] have presented a novel security technique. In this method, input time series from different sensors are compared using dynamic time warping (DTW) similarity in order to discover anomalies using a novelty detector that was previously trained with genuine, normal data as well as some realistic potential perception-layer attacks. They used the example of a smart cupboard (SC) with door magnetic sensors being subjected to magnetic field fluctuations in order to change how door events were perceived to exemplify this method. The experimental findings demonstrated the performance in detecting perception layer attacks without needlessly alerting the user during normal SC usage. More specifically, in the studies, 3.5% of the typical usages could not be authenticated, yet none of them were identified as attacks. Additionally, 95.5% of attacks on the perception layer were correctly detected, 4.20% of attacks were not classified, and only 0.30% of attacks were incorrectly labeled as common usages. Every one of these statistics was run on every analysis day. If they are not discovered on the first day, all attacks will likely be found in a few days. Figure 7.9 illustrates a block diagram for safeguarding a smart furniture item from perception-layer attacks.

To identify aberrant traffic and distinguish DDoS attacks from the flash crowd (FC), a unique detection and classification mechanism is introduced in [8] work. In general, both types of traffic share similar characteristics, although they can be distinguished from one another by a few crucial differences. To get the desired result, numerous steps are taken in this system, and the analysis data from the traffic analysis is then processed further. The number of packets, size of the payload, and inter-packet arrival time variations are the main factors taken into account. Ultimately, FC traffic is identified and distinguished from DDoS attacks using a Naive Bayesian model. Their proposed model is shown in Fig. 7.10. To confirm the system's performance, various simulations are created and compared with some existing methods. Their experimental and simulation findings demonstrate that their proposed detection system can distinguish DDoS attack traffic from FC with more than 93% accuracy (CAIDA–DDoS Attack 2007 and FIFA World Cup are two real-world datasets used in this research.)

7.6 CONCLUSIONS

Cybersecurity has become a global concern for establishing improved security mechanisms to investigate and react to cyberattacks. In this chapter, we identify several application and service domain vulnerabilities inherent to the IoT and smart systems. The ineffectiveness of conventional security solutions in detecting novel cyberattacks renders them insufficient. Numerous applications of cybersecurity systems make use of machine learning techniques. In this chapter, we've covered threats to IoT and smart systems, as well as a quick overview of malware detection and evasion approaches. It is essential to investigate novel cyberattacks while simultaneously building

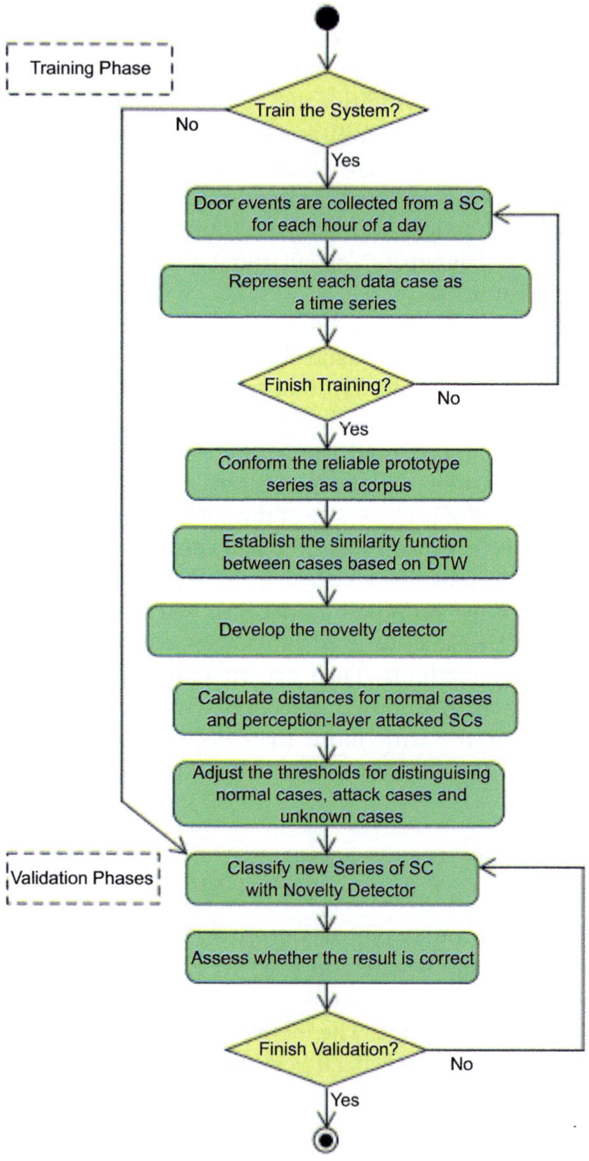

Fig. 7.9 Block diagram for securing an SC against magnetic perception-layer attacks [7]

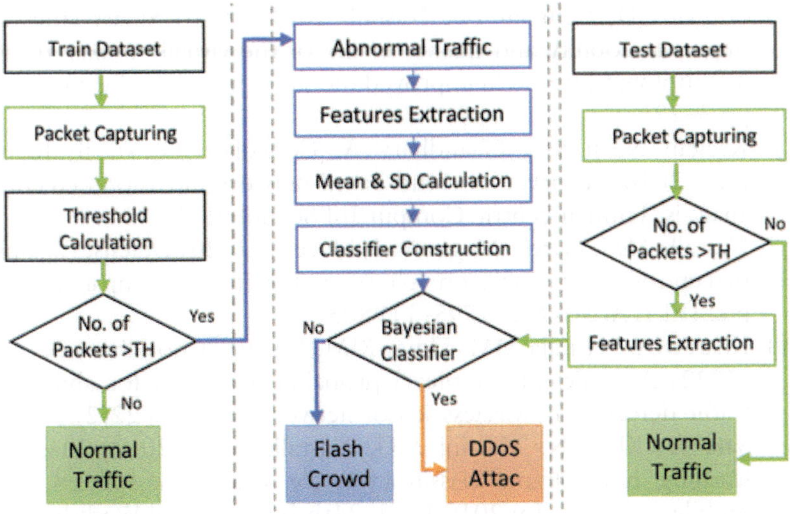

Fig. 7.10 Bayesian model to detect abnormal data traffic and discriminate DDoS attacks from FC [8]

and executing solutions to resist these cyberattacks, so the IoT and smart systems could be utilized to their full potential.

References

1. Nobakht M, Sivaraman V, Boreli R (2016) A host-based intrusion detection and mitigation framework for smart home IoT using openflow. In: 2016 11th International conference on availability, reliability and security (ARES). IEEE, pp 147–156
2. Nasralla MM (2021) Sustainable virtual reality patient rehabilitation systems with iot sensors using virtual smart cities. Sustainability 13(9):4716
3. Bendiab G, Shiaeles S, Alruban A, Kolokotronis N, IoT malware network traffic classification using visual representation and deep learning. In: 2020 6th IEEE Conference on Network Softwarization (NetSoft). IEEE, pp 444–449
4. Khattak SBA, Jia M, Marey M, Nasralla MM, Guo Q, Gu X (2022) A novel single anchor localization method for wireless sensors in 5G satellite-terrestrial network. Alexandria Eng J 61(7):5595–5606

5. Sobnath D, Rehman IU, Nasralla MM (2020) Smart cities to improve mobility and quality of life of the visually impaired. In: Technological trends in improved mobility of the visually impaired, pp 3–28
6. Keegan N, Ji S-Y, Chaudhary A, Concolato C, Yu B, Jeong DH (2016) A survey of cloud-based network intrusion detection analysis. Human-centric Comput Inf Sci 6(1):1–16
7. Nasralla MM, García-Magariño I, Lloret J (2020) Defenses against perception-layer attacks on IoT smart furniture for impaired people. IEEE Access 8:119795–119805
8. Khan MA, Nasralla MM, Umar MM, Khan S, Choudhury N et al (2022) An efficient multilevel probabilistic model for abnormal traffic detection in wireless sensor networks. Sensors 22(2):410
9. Saki H, Khan N, Martini MG, Nasralla MM (2019) Machine learning based frame classification for videos transmitted over mobile networks. In: 2019 IEEE 24th International Workshop on Computer Aided Modeling and Design of Communication Links and Networks (CAMAD). IEEE, pp 1–6
10. Sobnath D, Isiaq SO, Rehman IU, Nasralla MM (2020) Using machine learning advances to unravel patterns in subject areas and performances of university students with special educational needs and disabilities (MALSEND): a conceptual approach. In: Fourth International Congress on Information and Communication Technology. Springer, pp 509–517
11. Pachhala N, Jothilakshmi S, Battula BP (2021) A comprehensive survey on identification of malware types and malware classification using machine learning techniques. In: 2021 2nd International Conference on Smart Electronics and Communication (ICOSEC). IEEE, pp 1207–1214
12. Wurm J, Hoang K, Arias O, Sadeghi A-R, Jin Y (2016) Security analysis on consumer and industrial IoT devices. In: 2016 21st Asia and South Pacific design automation conference (ASP-DAC). IEEE, pp 519–524
13. Aman MN, Chua KC, Sikdar B (2017) A light-weight mutual authentication protocol for IoT systems. In: GLOBECOM 2017-2017 IEEE Global Communications Conference. IEEE, pp 1–6

14. Sengupta J, Ruj S, Bit SD (2020) A comprehensive survey on attacks, security issues and blockchain solutions for IoT and IIot. J Netw Comput Appl 149:102481
15. Khanmohammadi K, Ebrahimi N, Hamou-Lhadj A, Khoury R (2019) Empirical study of android repackaged applications. Empir Softw Eng 24(6):3587–3629
16. Yang Z, Chao F, Chen X, Jin S, Sun L, Du X (2022) DroidFDR: automatic classification of android malware using model checking. Electronics 11(11):1798
17. Yadav CS, Sharan A (2018) Automatic text document summarization using graph based centrality measures on lexical network. Int J Inf Retr Res (IJIRR) 8(3):14–32
18. Yadav CS, Sharan A, Joshi ML (2014) Semantic graph based approach for text mining. In: 2014 International Conference on Issues and Challenges in Intelligent Computing Techniques (ICICT). IEEE, pp 596–601
19. Yang Y, Du X, Yang Z, Liu X (2021) Android malware detection based on structural features of the function call graph. Electronics 10(2):186
20. Guebli W, Belkhir A (2021) Inconsistency detection-based LOD in smart homes. Int J Semant Web Inf Syst (IJSWIS) 17(4):56–75
21. Kouliaridis V, Barmpatsalou K, Kambourakis G, Chen S (2020) A survey on mobile malware detection techniques. IEICE Trans Inf Syst 103(2):204–211
22. Khanna N, Sachdeva M (2019) A comprehensive taxonomy of schemes to detect and mitigate blackhole attack and its variants in MANETs. Comput Sci Rev 32:24–44
23. Cid-Fuentes JÁ, Szabo C, Falkner K (2018) An adaptive framework for the detection of novel botnets. Comput Secur 79:148–161
24. Ngo Q-D, Nguyen H-T, Le V-H, Nguyen D-H (2020) A survey of IoT malware and detection methods based on static features. ICT Express 6(4):280–286
25. Han W, Xue J, Wang Y, Liu Z, Kong Z (2019) MalInsight: a systematic profiling based malware detection framework. J Netw Comput Appl 125:236–250

26. Wang S, Chen Z, Yan Q, Yang B, Peng L, Jia Z (2019) A mobile malware detection method using behavior features in network traffic. J Netw Comput Appl 133:15–25
27. Noor U, Anwar Z, Malik AW, Khan S, Saleem S, A machine learning framework for investigating data breaches based on semantic analysis of adversary's attack patterns in threat intelligence repositories. Futur Gener Comput Syst 95:467–487
28. Taheri R, Ghahramani M, Javidan R, Shojafar M, Pooranian Z, Conti M (2020) Similarity-based android malware detection using hamming distance of static binary features. Futur Gener Comput Syst 105:230–247
29. Alazab M, Alazab M, Shalaginov A, Mesleh A, Awajan A (2020) Intelligent mobile malware detection using permission requests and API calls. Futur Gener Comput Syst 107:509–521
30. Borchani Y (2020) Advanced malicious beaconing detection through AI. Netw Secur 2020(3):8–14
31. Visu P, Lakshmanan L, Murugananthan V, Cruz MV (2019) Software-defined forensic framework for malware disaster management in internet of thing devices for extreme surveillance. Comput Commun 147:14–20
32. Yonamine S, Taenaka Y, Kadobayashi Y (2022) Tamer: a sandbox for facilitating and automating IoT malware analysis with techniques to elicit malicious behavior. In: ICISSP, pp 677–687
33. Kachare GP, Choudhary G, Shandilya SK, Sihag V (2022) Sandbox environment for real time malware analysis of IoT devices. In: International Conference on Computing Science, Communication and Security. Springer, pp 169–183
34. Noor M, Abbas H, Shahid WB Countering cyber threats for industrial applications: an automated approach for malware evasion detection and analysis. J Netw Comput Appl 103:249–261
35. Fuji R, Usuzaki S, Aburada K, Yamaba H, Katayama T, Park M, Shiratori N, Okazaki N (2019) Blockchain-based malware detection method using shared signatures of suspected malware files. In: International Conference on Network-Based Information Systems. Springer, pp 305–316

36. Jian Y, Kuang H, Ren C, Ma Z, Wang H (2021) A novel framework for image-based malware detection with a deep neural network. Comput Secur 109:102400
37. d'Estalenx A, Gañán C (2021) Nurse: end-user IoT malware detection tool for smart homes. In: 11th International Conference on the Internet of Things, pp 134–142
38. Ullah F, Naeem H, Jabbar S, Khalid S, Latif MA, Al-Turjman F, Mostarda L (2019) Cyber security threats detection in internet of things using deep learning approach. IEEE Access 7:124379–124389
39. Andrade RO, Yoo SG, Tello-Oquendo L, Ortiz-Garcés I (2020) A comprehensive study of the IoT cybersecurity in smart cities. IEEE Access 8:228922–228941
40. Laszka A, Potteiger B, Vor obeychik Y, Amin S, Koutsoukos X (2016) Vulnerability of transportation networks to traffic-signal tampering. In: 2016 ACM/IEEE 7th International Conference on Cyber-Physical Systems (ICCPS). IEEE, pp 1–10
41. Bagga P, Das AK, Wazid M, Rodrigues JJ, Park Y (2020) Authentication protocols in internet of vehicles: taxonomy, analysis, and challenges. IEEE Access 8:54314–54344
42. Soltan S, Yannakakis M, Zussman G, React to cyber attacks on power grids. IEEE Trans Netw Sci Eng 6(3):459–473
43. Taormina R, Galelli S, Tippenhauer NO, Salomons E, Ostfeld A (2017) Characterizing cyber-physical attacks on water distribution systems. J Water Resour Plan Manag 143(5):04017009
44. Butun I, Österberg P, Song H (2019) Security of the internet of things: vulnerabilities, attacks, and countermeasures. IEEE Commun Surv Tutor 22(1):616–644
45. Hassija V, Chamola V, Saxena V, Jain D, Goyal P, Sikdar B (2019) A survey on IoT security: application areas, security threats, and solution architectures. IEEE Access 7:82721–82743
46. Kankanhalli A, Charalabidis Y, Mellouli S (2019.) IoT and AI for smart government: a research agenda. Gov Inf Q 36(2):304–309
47. Bailey D, Coleman Y (2018) Urban IoT and AI: how can cities successfully leverage this synergy? Retrieved Feb, vol 23, p 2019

48. Ozer M, Varlioglu S, Gonen B, Bastug M (2019) A prevention and a traction system for ransomware attacks. In: 2019 International Conference on Computational Science and Computational Intelligence (CSCI). IEEE, pp 150–154
49. Bayer U, Moser A, Kruegel C, Kirda E (2006) Dynamic analysis of malicious code. J Comput Virol 2(1):67–77
50. Ye Y, Li T, Adjeroh D, Iyengar SS (2017) A survey on malware detection using data mining techniques. ACM Comput Surv (CSUR) 50(3):1–40
51. Milosevic J, Sklavos N, Koutsikou K (2016) Malware in IoT software and hardware
52. Milosevic J, Regazzoni F, Malek M (2017) Malware threats and solutions for trustworthy mobile systems design. In: Hardware security and trust. Springer, pp 149–167
53. Wazid M, Das AK, Rodrigues JJ, Shetty S, Park Y (2019) IoMT malware detection approaches: analysis and research challenges. IEEE Access 7:182459–182476
54. Uchenna CC, Jamil N, Ismail R, Yan LK, Mohamed MA (2021) Malware threat analysis techniques and approaches for iot applications: a review. Bull Electr Eng Inf 10(3):1558–1571
55. Ahamed J, Rajan AV (2016) Internet of things (IoT): application systems and security vulnerabilities. In: 2016 5th International Conference on Electronic Devices, Systems and Applications (ICEDSA). IEEE, pp 1–5
56. Clincy V, Shahriar H (2019) IoT malware analysis. In: 2019 IEEE 43rd Annual Computer Software and Applications Conference (COMPSAC), vol 1. IEEE, pp 920–921
57. Aslan ÖA, Samet R (2020) A comprehensive review on malware detection approaches. IEEE Access 8:6249–6271
58. Cui L, Xie G, Qu Y, Gao L, Yang Y (2018) Security and privacy in smart cities: Challenges and opportunities. IEEE Access 6:46134–46145
59. Adkins F, Jones L, Carlisle M, Upchurch J (2013) Heuristic malware detection via basic block comparison. In: 2013 8th International Conference on Malicious and Unwanted Software: The Americas (MALWARE). IEEE, pp 11–18

60. Barriga JJ, Yoo SG (2017) Malware detection and evasion with machine learning techniques: a survey. Int J Appl Eng Res 12(18):7207–7214
61. Shaukat K, Luo S, Varadharajan V, Hameed IA, Xu M (2020) A survey on machine learning techniques for cyber security in the last decade. IEEE Access 8:222310–222354
62. Patel C, Vyas S, Saikia P et al (2022) A futuristic survey on learning techniques for internet of things (IoT) security: developments, applications, and challenges.
63. Chen Z, Liu J, Shen Y, Simsek M, Kantarci B, Mouftah HT, Djukic P (2022) Machine learning-enabled IoT security: open issues and challenges under advanced persistent threats. ACM Comput Surv (CSUR) 55(5):1–37

CHAPTER 8

A Multiclass Classification Approach for IoT Intrusion Detection Based on Feature Selection and Oversampling

Zayna Amierh, Lina Hammad, Raneem Qaddoura, Huthaifa Al-Omari, and Hossam Faris

8.1 Introduction

The concept Internet of Things (IoT) revolves around a time when there will be more objects linked to the Internet than there will be humans. Under the current Internet infrastructure, the Internet of Things refers to

Z. Amierh • L. Hammad • R. Qaddoura (✉) • H. Al-Omari
School of Computing and Informatics, Al Hussein Technical University, Amman, Jordan
e-mail: Zayna.amierh@htu.edu.jo; Lina.Hammad@htu.edu.jo; raneem.qaddoura@htu.edu.jo; Huthaifa.AlOmari@htu.edu.jo

H. Faris
Altibbi, Amman, Jordan

Information Technology Department, The University of Jordan, Amman, Jordan

Research Centre for Information and Communications Technologies of the University of Granada (CITIC-UGR), University of Granada, Granada, Spain
e-mail: hossam.faris@ju.edu.jo

I. Almomani et al. (eds.), *Cyber Malware*, Security Informatics and Law Enforcement, https://doi.org/10.1007/978-3-031-34969-0_8

the interconnection of recognizable embedded computing devices [1]. The simple idea behind the Internet of Things is to enable autonomous and safe data exchange and connection between applications and real-world devices in addition to connecting the physical and virtual worlds through personal devices, servers, smart cities, and anything connected to the Internet [2]. IoT is the next evolution of the Internet, and its effect on education, communication, enterprise, research, government, and industry has made it one of humanity's most significant and influential creations [1].

The Internet of Things has advanced exponentially in recent years. Yet, IoT is not a new concept; it was initially presented in 1999 by Kevin Ashton, co-founder and executive director of the Auto-ID Center, who suggested that computers, and hence the Internet, are completely reliant on human understanding [3]. Despite this, in 2003, IoT did not exist yet due to the limited number of connected things and also since omnipresent devices such as smartphones were just being introduced. With time that has passed by, IoT came out of the shadows in 2009 when actual implementation started [4].

IoT has been applied in many fields that we would not have expected such as in agriculture and predicting the occurrence of natural calamities [5]. Furthermore, the Internet of Things has been used in health monitoring systems [6], surveillance monitoring systems [7], autonomous vehicles [8], smart cities [9], and a variety of other applications where all information can be used. With the help of Big Data, IoT has now become extremely powerful, allowing us to collect and analyze large amounts of data in a variety of ways, assisting in the transformation of businesses, industries, government services, and people's lives [10].

As the number of Internet-connected devices and new IoT applications increases, security threats in each device/network develop as a result of network intrusions attacks that can occur as a result of various security vulnerabilities that allow this, as well as IoT devices that do not recognize and consider all security flaws [11]. As an example, in a survey on real IoT devices having security flaws, commercial "smart" services and products (smart appliances, smart watches, smart TVs, and so on) are provided with insufficient, incomplete, and ill-designed security mechanisms, resulting in numerous risks relating to access to sensitive information or critical controls [12]. Having said that, intrusion detection systems (IDS) are a security mechanism that can detect anomalies and malicious activities in a network and protect against three types of attacks: anomaly-based attacks, signature-based attacks, and even a mixture known as hybrid attacks. Signature-based techniques identify attacks based on their signatures. However, due to the evolution of intrusion attacks, this technique is not always capable of detecting zero-day or newly evolved attacks. Anomaly-based attacks, on the other hand, can attempt to detect attacks based on abnormal network behavior. In hybrid techniques, it is a mixture of both previous detection techniques [13]. Machine learning techniques (both unsupervised and supervised learning methods) have been used with these techniques to improve by automatically creating unique rules for signature-based IDS or adapting the detection patterns of anomaly-based IDS for creating predictions and analysis based on the data given to detect patterns [14].

With all the existing techniques for intrusion detection, a substantial amount of data is generated from IoT devices, and this can be quite an issue for the methods for detection in terms of collecting, storing, and processing the data in addition to the delay in prediction and actions [15]. Due to the production of all this data, it is known that it could be associated with an imbalance of this data. An imbalanced dataset has non-uniformly distributed instances with regard to their corresponding labels [16]. The problem of the imbalanced dataset in machine learning applications is the tendency to aggressively identify the instances with the minority labels as instances of the majority class, which degrades the quality of the machine learning outcomes [17]. This is a problem since it can compromise the real performance of the machine learning techniques and algorithms that are used in the detection and classification since most standard methods assume balanced class distribution and mis-classification costs as equal leading to inaccuracies and weak representations and results for the distributive characteristics [18]. Most machine learning techniques face challenging the imbalanced nature of the data generated especially in the multiclass problem found in the dataset. So far, this issue could be overcome with the many different solutions such as oversampling, undersampling, and cost-sensitive learning methods [19]. If not resolved, they can lead to a low predictive performance in detection and identification of normal activity vs intrusion to classify them based on normal and types of attacks [20]. For detection and classification in the IDS, many approaches were combined to gain different results and accuracy whether with the use of supervised or unsupervised learning. Classification based on multiclass problems is accompanied with oversampling techniques specific for the adjustment of the data distribution in the multiclass problem for detection and classification of the intrusions in spite of the change of percentages of the normal and attack traffic [21]. With the datasets used for the detection of intrusions, the multiclasses must have identified feature variables that could influence and play a big role in the prediction of this class to adopt valid test and evaluation to reflect the trends and evident diversity [22].

Putting all these stages together, it forms a pipeline for the classification of IoT intrusion detection with the use of feature selection, oversampling, and feature importance. Although this research has been adopted before, it has not been considered on one of the most recent imbalanced datasets using new popular classification techniques. Hence, in this chapter, the main contributions are as follows:

- Develop the multiclass classification to classify the category label for IoT intrusion attacks.
- Application of heuristic approach for feature selection to adopt valid predictions for detection of IoT intrusion attacks.
- Application of oversampling technique to use for the multiclass dataset problem to solve imbalance distribution.
- Specify the influencing feature variables that play a big role in the prediction power.
- Application of pipeline on recent dataset IoTID2020 using new popular algorithms.

The rest of this chapter is organized by discussing the literature review, security system framework, background, research methodology, experimental results, and discussions along with the conclusion and future works of this research.

8.2 Literature Review

As IoT is evolving drastically over the years, it can be seen that a massive amount of data is being produced by all the IoT devices and it is very hard to detect the intrusion activities using proper mechanism even with the use of machine learning techniques (whether supervised or unsupervised learning techniques). To detect the intrusions and normal activities for classification, Qaddoura et al. [23] proposed an entire approach in multi-stages for the classification based on clustering, performing reduction along with oversampling and classification methods, as well as working on the IoT training data with undersampling, while the dataset is still left representative for training by using the unique reduction and clustering approach. Oversampling was also done to solve the imbalanced distribution of the data in the classes. With all the procedures done, it was experimented on the IoTID2020 dataset by dividing the training data into three clusters with the use of k-means clustering which were then reduced by 10% before aggregating them to the three reduced clusters. SVM-SMOTE was furtherly used with an oversampling ratio of 0.9 and then aggregated into an enlarged one for supplementarily producing the oversampled data classification model using the single-hidden layer feed-forward neural network (SLFN) classification method. When evaluated, it was shown to be exceeding other approaches based on the G-mean (GM), precision (PREC), accuracy (ACC), and recall (REC) [23]. The same authors also proposed a multi-layer approach for the IoT intrusion detection using the IoTID2020 dataset to predict the intrusion identification and the category label using SLFN and long short-term memory (LSTM) with oversampling [17].

With the ongoing research on detecting IoT intrusions, there are many aspects and pathways to follow with many theories and techniques to investigate. Finding approaches for the classification of IoT intrusions can be quite challenging especially due to the many issues and obstacles surrounding it. For example, it has been studied that class imbalance is a big type of issue in classification since there are classes that are marginalized when compared to others. This raises an effectiveness conflict particularly in minority class prediction when using algorithms of machine learning. There are different numerous approaches to tackling this matter although the majority focus on bi-class scenarios in the imbalance problems. Therefore, it has been proved that dealing with multiclass problems based on these algorithms is less efficient and has negative consequences. With this said, Wang and Yao [21] have addressed this point by considering why addressing multiclass problems tends to be difficult using these approaches. It was concluded unsatisfactory of strategizing the effect of the multiclass on the random and undersampling execution processes in the multi-majority and multi-minority class cases. Due to this, they proposed their developed ensemble algorithm named AdaBoost. NC [21] along with oversampling to resolve the multiclass problem and improve the balance and recognition of minority classes that can improve the performance in classification [21].

Furthermore, Abdi and Hashemi [20] worked on opposing the multiclass imbalance problem using Mahalanobis distance-based oversampling technique (MDO) to minimize the consequential challenge of the overlapping risk that can occur between regions of different classes in the detection of IoT intrusions using model-based solutions such as ensemble learning to address these issues. With the application of the simple oversampling technique MDO, it overcomes the multiclass problems by considering each class candidate and oversampling the minority class and inventing a synthesis instance that can be useful for maintaining the covariance data structure for the minority classes that can help the prevalent issue of overlaps. The MDO strategy was compared with many existing oversampling techniques and tested across 20 multiclass UCI and KEEL benchmark datasets with a few classifiers that are different. It shows that this technique works great in multiclass imbalance problems due to higher capability of learning the boundaries of different classes such as feature space similarities that are hard to learn by other methods. Additionally, not only has the MDO reduced the oversampling risk by sampling the minority classes and creating samples in dense feature space areas, but it has also outperformed other data-level solution algorithms based on the precision of minority classes and MAUC in classification [20].

However, the MDO [20] has demonstrated that it is only applicable to numerical attributes in datasets and can ruin the majority class accuracy due to the generation of unrealistic samples and immoderate synthetic samples for the minority classes. Thus, Yang et al. [24] put forward the extension of the study and adapted it to propose the adaptive Mahalanobis distance-based oversampling (AMDO) [24] which handles not only the mixed-type attributes in multiclass imbalanced datasets effectively but also improves the MDO performance using a partially stabilized re-sampling method using a strategy to gain the adaptive samples. Based on comparison and testing, AMDO shows outperforms MDO in terms of numeric datasets and mixed-typed datasets and concluded higher accuracy of the minority classes and classifiers performance in most datasets as promising results of precision and AUC [24].

As many other new techniques arise to detect IoT intrusions, many of these machine learning models assume that each of the classes to detect from contains an equal number of samples. Yet, with the imbalance data nature in IoT security, a very poor performance of predicting and identifying anomalies is observed. This has made it very difficult to attempt to design a model for detecting anomalies using the machine learning models. Based on this, Dash et al. [15] initiated a different technique for anomaly prediction with a synthetic minority oversampling technique (SMOTE) with a multiclass adaptive boosting ensemble learning-based model to be tested on DSOS data in comparison and other machine learning approaches in handling this issue. It was concluded by the authors [15] that the imbalanced multiclass nature was handled successfully and effectively in identifying the normal activities vs abnormal activity along with the anomaly types. Moreover, it surely shows that based on evaluation metrics and performance indexes, greater efficiency was achieved in contrast to other approaches [15].

Finding a procedure to detect and classify network attacks in multiclass problems is used in intrusion detection systems (IDS), and they use machine learning models to handle whether based on anomaly behavior or signatures that are known to the systems. Many techniques could be used for detection whether based on patterns or rules for multiclass problems. Alaiz-Moreton et al. [14] did research on three methods in machine learning to decide on the normal vs abnormal activity within the same time period. Many modes such as LSTM, GRU, and XGBoost were used. Models based on recurrent neural networks have high accuracy due to their time and sequence importance in the attacks, while the XGBoost was used for its beneficial structure. Based on usage, ensemble techniques had the best results overall, followed by deep learning techniques, having the worst results for linear models [14].

Furtherly, knowing there are many techniques for classification based on rules and patterns of the attacks knowledge, it is quite prone to errors in the intrusion detection system due to the presence of the class-imbalance data. Owing to this, Panigrahi et al. [25] looked into this issue and introduced a C4.5-based detector in the system with the consolidated tree construction algorithm along with Supervised Relative Random Sampling (SRRS) and multiclass feature selection for efficient intrusion detection. This system was evaluated on the NSL-KDD dataset and the CICIDS2017 dataset using 34 features showing the accuracy of 99.96% and 99.95% [25].

Although this approach discussed by Panigrahi et al. [25] could be a solution. Yet, many other researchers have other techniques to suggest. Iwendi et al. [26] proposed an improvement to the system for bi-class and multiclass data by the use of CFS + Ensemble Classifiers (Bagging and Adaboost) with base classifiers (J48, RandomForest, and Reptree). This was tested on KDD99 and NSL-KDD datasets for binary and multiclass classification resulting in 99.90% and 98.60% detection rates and 0 and 0.5% false alarm rates correspondingly [26].

In the classification of intrusion attacks, a large quantity of data is processed, and the detection rate is quite low as a default especially if data is highly imbalanced. Thus, Miah et al. [27] proposed a multi-layer classification approach using cluster-based undersampling with a random forest classifier to improve the detection rate to classify minority-class intrusions addressing imbalanced and overfitting problems. The KDD99 dataset was used in the experiment showing that the proposed method raises detection rates and abates false-positive rates compared to other classifiers with 87% achievement as other algorithms can achieve less than 30% [27].

Due to the rise of applications of IoT, they have become more subjected to attacks that are malicious, and now most mechanisms are unable to overcome and protect against completely even if they try to defend against the majority types of attacks, despite that the systems for detecting traditionally have many flaws in their detection and time efficiency. Yet, they can identify suspicious attacks based on the behavior which is abnormal in the IoT devices. Now timely response and implementation of measure to protect effecting is essential and is done based on the collecting of data from the network to view the behavior although the data contains many features, featural dimensions that are high with complex structures to deal with, so this is an issue for the performance of detecting in the existing algorithms used. This was addressed by Zheng et al. [28] that came up with a better version detection algorithm based on the LDA ELM classification to fulfill efficient and timely detection by the use of improving the linear discriminant analysis LDA and adding similarity functions that are special to achieve better spatial separation after the reduction of the dimensions of the data. In addition to this, the paper used the extreme learning machine (ELM) classifier with this speedy classification for better decisions, with further experimentation and processes that managed to test this on the NSL-KDD dataset and compare it against other algorithms concluding the highest accuracy and detection rate [28].

All in all, it can be viewed that there are many different ways researchers view and think about solving these problems with various approaches using different algorithms and techniques accompanied by considering different factors. Based on the different views, each concludes different results in their papers giving a bigger reason to work on this topic further since comparing the results together and implementing different multiple methodologies to classify intrusion attacks in multiclass problems can propose a whole new hypothesis to investigate and build on next, especially in IoT.

8.3 Security System Framework

The intrusion detection prevention system (IDPS) is a mechanism that focuses on the security of the network layer of an IoT system. The way this system works is by analyzing the packets of data and generating responses in real time with fast responses and high-volume data processing capabilities. In the analysis process, the data is collected into a monitored environment and preprocessed. Beyond this, it is fed through the classification model to predict whether the data from the network attack matches an intrusion attack or not. If it is a normal packet, no alarm is given, and the packet passes through the network. Yet, if the data is malicious, it predicts which type of attack this packet matches (e.g., DoS, Mirai, MITM, Scan) and alerts the system to take action to prevent potential harm to the IoT network [29]. An interpretation of this system framework is shown in Fig. 8.1.

8.4 Background

This section discusses the preliminary information needed to understand the main parts of the proposed methodology. It mainly includes a discussion on the XGBoost and CatBoost classifier algorithms.

8.4.1 XGBoost

XGBoost algorithm, like many other ensemble learning algorithms, is used for regression and classification for supervised learning problems and large datasets where there is multiple features in the training data to predict a target variable. This algorithm was developed by Chen and Guestrin [30] and was optimized with the structure of gradient boosting. The XGBoost is a regression tree that is popular for its scalability in all scenarios as it can allow the system to run ten times faster than any solution on a single machine. This is due to its algorithmic optimizations as sparse data is handled by a novel tree learning procedure; instance weights in approximation tree learning is handled by a theoretically justified weighted quantile sketch procedure, and this helps in the split finding algorithms [31]. Learning is sped up with the use of parallel and distributed computing. This helps in solving complex problems in machine learning allowing for more rapid model generation in a fast and accurate way [32]. It works by integrating the estimates of several simpler, weaker models to try to accurately predict

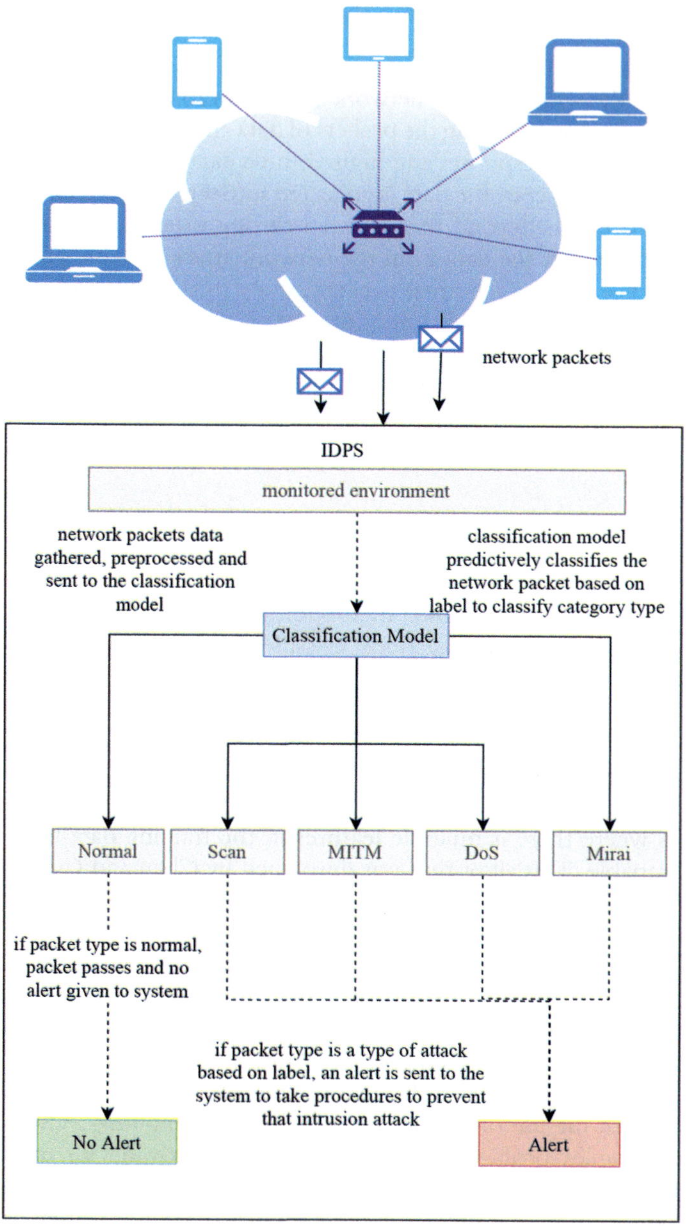

Fig. 8.1 Security system framework to classify and alert IoT intrusion attacks

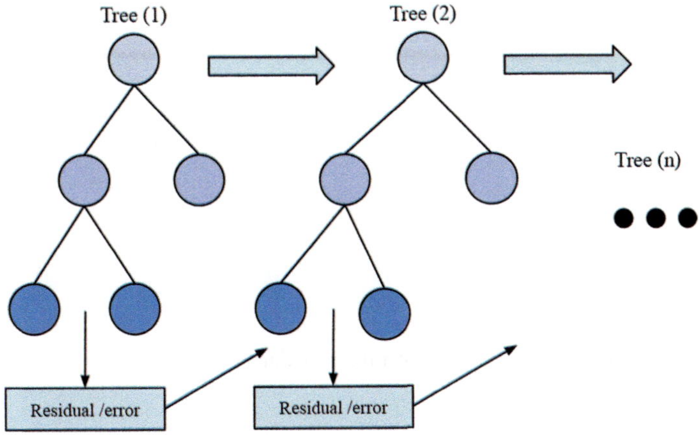

Fig. 8.2 Representation of decision trees in XGBoost

a target variable. A convex loss function (determined from the anticipated and target outputs difference) is combined with a penalty term in XGBoost for model complexity to diminish a regularized regression tree function (objective function). The training process is repetitious, with new trees being included that forecast the errors or residuals of prior trees, which are then integrated with previous trees to provide the final prediction as shown in Fig. 8.2. Gradient boosting in the XGBoost uses a gradient descent approach to reduce loss when new models are added. Yet, XGBoost cannot handle categorical features so one-hot-encoding must be used to transform the categorical features into numerical values for the algorithm. As a whole, the XGBoost algorithm is used as it is a recent and popular strong algorithm that supports numerous objective functions, such as regression, classification, and ranking with a fast execution speed and a highly accurate model performance [33].

8.4.2 CatBoost

CatBoost is a depth-wise gradient boosting algorithm that was developed by Yandex. Within gradient boosting, the trees are made one after the present one where the previous trees can't be altered, but the results are used to improve the next trees. The CatBoost algorithm uses past

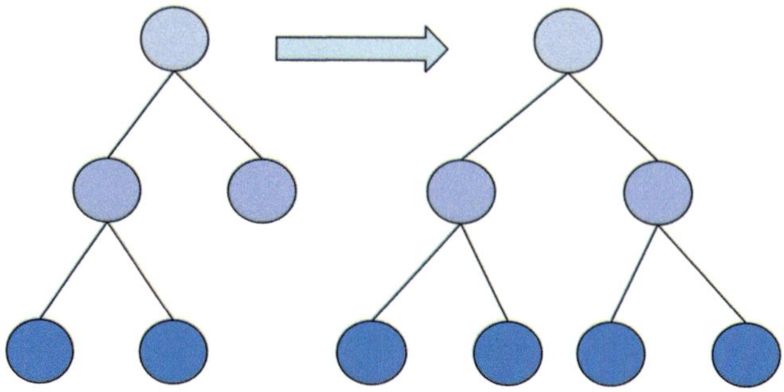

Fig. 8.3 Representation of decision trees in CatBoost

decision trees to grow a balanced tree where the left and right splits for each level of the tree are made from the same features as shown in Fig. 8.3. This algorithm can handle categorical features and numerical values and reduce overfitting with very less prediction time at a high accuracy giving the advantage that it can be used in complex problems with large datasets for classification and regression. CatBoost has the flexibility of giving categorical columns indices so one-got encoding can be used using one-hot-max-size. The CatBoost utilizes an encoding method to reduce the overfitting by permuting the set of inputs in an irregular order and converting label values from floating point or category to integer in addition to transforming the categorical feature values to numeric values using the formula:

$$Avg\_target = \frac{countInClass + prior}{totalCount + 1} \tag{8.1}$$

where the countInClass is number of times the label value was equal to 1 for the present categorical feature value objects. Prior is the preliminary value for the numerator and is determined by the parameters at the start. The total count is the objects total number to the current that has the categorical value feature matching. Furthermore, minimal variance sample (MVS) is a stochastic gradient boosting weighted sampling form that CatBoost employs. Weighted sampling happens at the tree level instead of the split level in this technique. Each boosting tree's observations are sampled in such a way that split scoring accuracy is maximized. The reason why CatBoost is popular and in demand for usage is that great results are provided with the default parameters; thus, less time is needed for tuning parameters, and it reduces overfitting due to improved accuracy, and the usage of the CatBoost model applier allows fast prediction [34].

8.5 Methodology

In the multiclass perspective approach for detecting intrusion attacks in this research, the primary quantitative research method is used as a model for the experiments as it is developed to analyze specific data from a dataset by using sets of variables setting one as constant and measuring the differences against the other (training and testing sets), and this means that the information is gathered through the self-conducted research methods. The secondary research was used in the literature review, and the information was taken from different studies. This supports the proposed research. Moreover, as quantitative research has other approaches such as using surveys, they are not used, and neither is the qualitative research (e.g., interviews) be used as they are not necessary for this project and do not meet our objectives. The reason behind it is that both mentioned techniques focus on human experiences, behavior, and opinions, while this experiment looks at the home network's traffic and corresponding connected devices for malicious activities, and this is not associated with any known ethical issues due to no human participation in research as the data is readily available online with no human data [35]. Thus, the quantitative experimental research approach is the most effective to reach the objectives. Yet, caution must be taken upon working on quantitative research as it involves limitations and drawbacks of difficulty in understanding the context of the phenomenon and explaining complex issues due to data not being robust enough and requires time and cost which is expensive [36].

Furthermore, in this methodology, the main focus is on the following aspects as shown in Fig. 8.4:

- Problem understanding and formulation
- Data collection
- Data preprocessing
- Model development
- Evaluation and assessment

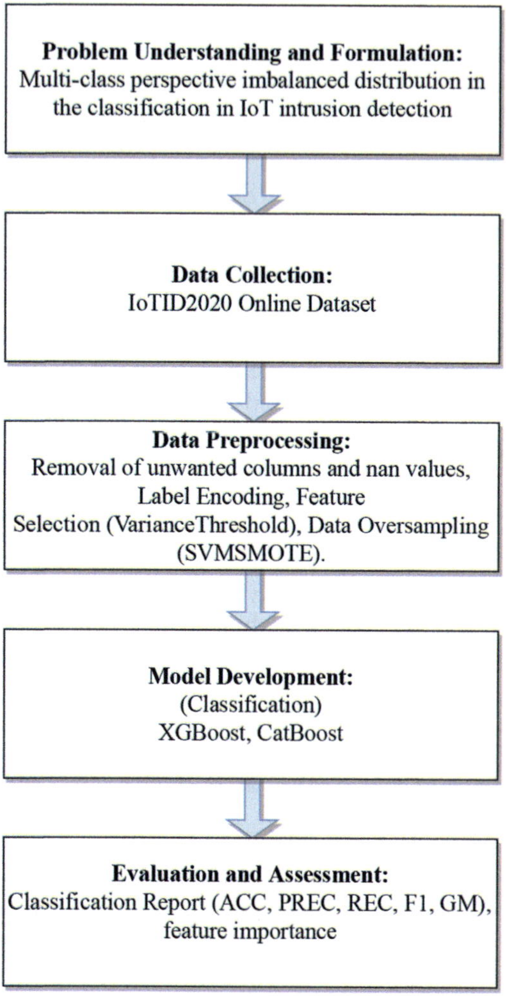

Fig. 8.4 Multiclass classification research methodology

8.5.1 Problem Understanding and Formulation

Intrusion detection and classification have risen in importance as the use and development of IoT devices and applications have increased. As technology advances, it creates new vulnerabilities and weaknesses in the system, network, and device itself. With all of the vulnerabilities, more network intrusion attacks can target these vulnerabilities, causing adverse consequences.

With the intrusion detection methods available today, detection algorithms are required to detect all types of attacks. Because the devices generate a large amount of data, not all machine learning algorithms are capable of detecting accurately and having less prediction time. Furthermore, the data collected may have an imbalanced distribution and contain multiclasses, therefore less likely and accuracy in detection of the attacks. As a result, a solution must be found. Many researchers have addressed this issue on old datasets yet using the basic known classifiers such as random forest, decision trees, logistic regression, KNN, and others. Due to this, a multiclass classification is implemented with feature selection and oversampling on a recent imbalanced dataset using recent popular classifiers.

8.5.2 Data Collection

To bring the research to light, the data needed for the experiment is completely taken from an online dataset called IoTID2020 [37] that includes 80 features and 625,783 instances. This dataset is a recent and imbalanced dataset that contains extensive network and flow-based features based on home environments with three main labels, the intrusion identification label, the category label, and the subcategory label (Shown in Table 8.1). The category label contains five different categories each with a different number of instances (Shown in Table 8.2). This dataset was also used in different new studies such as Qaddoura et al. [23] paper during the experiment implementation.

This data was obtained from a smart home environment that has a combination of IoT devices and interconnecting structures consisting of smart home device SKTNGU and EZVIZ Wi-Fi camera connected to the smart home Wi-Fi-router as it is connected to devices such as laptops, tablets, smartphones, and others. The data for IoTID2020 dataset was obtained as a result [38]. Although this dataset has the limitation of taking from the smart home environment, it is enough to test the methodology and gain results where it can be further improved and tested on other datasets.

Table 8.1 IoTID2020 taxonomy in terms of binary, category, and subcategory labels

Label	*Category*	*Subcategory*
Normal	Normal	Normal
Anomaly	DoS	Synflooding
	Mirai	Brute Force
		HTTP Flooding
		UDP Flooding
	MITM	ARP Spoofing
	Scan	Host Port
		OS

Table 8.2 Distribution of data instances based on category label

Category	*Number of instances*
DoS	59,391
Mirai	415,677
MITM	35,377
Scan	75,265
Normal	40,073
Total	625,783 instances

8.5.3 Data Preprocessing

The fundamental step after collecting the data is to manipulate and transform the data to increase the quality and enhance the performance of the experiment. To do so, unwanted columns shall be removed from the dataset along with nan values. The target column category is encoded using the label encoder to transform the column data into a numeric form to convert them into the machine-readable form [39]. Beyond this, the dataset is separated into features and the label where the label contains the feature category while the features contain all other column features. These are then used to split the data by 50% into training and testing set to be used in the rest of the procedure. Beyond this, feature selection using the variance threshold technique is implemented to remove the features with low variances (features with the same values in all samples) that does not meet the threshold (Shown in Table 8.3). This is done to decrease training time and reduce the risk of overfitting. Oversampling to the minority classes is next applied to the training set using the SVM-SMOTE to address the imbalance distribution of the classes of the dataset by using the SVM algorithm to detect samples to use for new synthetic samples generation. The data preprocessing stage can need quite a heavy capacity in terms of time and power, but it is essential to ensure the quality of the data for the experiment.

Table 8.3 Number of features selected and not selected based on the variance threshold technique

Feature selection (overall 80)/0.8 variance threshold	
# of features selected	# of features not selected
60	20

Table 8.4 Representation of metric evaluation

		Actual	
		Positive	Negative
Predicted	Positive	TP	FP
	Negative	FN	TN

Table 8.5 XGBoost experiment algorithm

Algorithm: SVM-SMOTE-XGBoost-classification
Input: dataset
Output: ACC, PREC, REC, GM, F1
1 train, test = split(dataset)
2 feature-selection-dataset= variance threshold
3 oversampled-dataset = SVM-SMOTE(feature-selection-dataset)
4 model = XGBoost(oversampled-dataset)
5 predicted-labels = predict(model, test)
6 ACC, PREC, REC, GM, F1 = evaluate(predicted-labels)
7 feature-importance = shap.summary.plot(model)

8.5.4 *Model Development*

After data preprocessing, the data is then passed into different classification models. The testing set of the data is the input for the model, which results in producing classified instances into the categories (DoS, MITM ARP Spoofing, Mirai, Scan, Normal). The main objective of classification is to build a model from items categorized to classify instances as correctly as possible by having many True positives (TPs) and True negatives (TNs) presented in Table 8.4. Such that data is fed into the models representing different classifiers specifically XGBoost and CatBoost since they are recent powerful algorithms with high prediction accuracy in short prediction timing that have strong potential in intrusion detection classification. Each classifier predicts the instances category as accurately as possible. It is essential to have high accuracy models to detect new changing attacks since most existing machine learning models are unable to identify some new attack patterns. Tables 8.5 and 8.6 refer to the algorithms for XGBoost and Catboost for model development, respectively.

Table 8.6 CatBoost experiment algorithm

Algorithm: SVM-SMOTE-CatBoost-classification
Input: dataset
Output: ACC, PREC, REC, GM, F1
1 train, test = split(dataset)
2 feature-selection-dataset= variance threshold
3 oversampled-dataset = SVM-SMOTE(feature-selection-dataset)
4 model = CatBoost(oversampled-dataset)
5 predicted-labels = predict(model, test)
6 ACC, PREC, REC, GM, F1 = evaluate(predicted-labels)
7 feature-importance = shap.summary.plot(model)

8.5.5 *Evaluation and Assessment*

The evaluation and analysis of the results are done by the following points.

- Measure the performance by calculating metrics GM, PREC, ACC, REC, and F1-score (F1) for each model using the Equations (8.2), (8.3), (8.4), (8.5), and (8.6) [23, 40].
- Compare XGBoost and CatBoost classifiers with random forest classifier, KNeighbors classifier, GaussianNB, logistic regression, decision tree classifier.

$$Accuracy = \frac{TP + TN}{TP + TN + FP + FN} \tag{8.2}$$

$$Precision = \frac{TP}{TP + FP} \tag{8.3}$$

$$Recall = \frac{TP}{TP + FN} \tag{8.4}$$

$$F - measure = \frac{2Precision \times Recall}{Precision + Recall} \tag{8.5}$$

$$G - Mean = \sqrt[n]{x_1, x_2, \ldots x_n} \tag{8.6}$$

The experiment is repeated three times where each has a scenario as follows:

- Scenario (1): Classification using XGBoost and CatBoost without data preprocessing.
- Scenario (2): Classification using XGBoost and CatBoost with feature selection.
- Scenario (3): Classification using XGBoost and CatBoost with feature selection and oversampling.

8.5.6 Post Analysis

To improve the predictive model performance and simplify and speed up the modeling process, feature importance is an essential technique to apply to the experiment. Feature importance techniques assign a specific score to each of the input features based on how useful they are in the prediction and classification of the target variables, and this as a result gives very useful insights about the data giving a better understanding of which features are relevant (highest scores) and irrelevant (lowest scores), thus reducing the number of input features to increase predictive model performance. Feature importance is used in the experiment for the analysis of which features are the most relevant in the prediction of the target variable in the XGBoost and CatBoost models. This is applied by finding the top important features using a different technique for each model [41].

XGBoost Feature Importance In the XGBoost, the general way to calculate the feature importance is with the boosted trees that are constructed and are used to extract each feature attribute importance scores based on the indication of how valuable each feature is in the boosted decision trees construction in the model. The more the feature has been used to make key decisions, the higher its relative importance [41]. Yet, another technique is called Shap which is model-agnostic and uses the Shapley values to estimate how does each feature contribute to the prediction for overall prediction [42].

CatBoost Feature Importance In the CatBoost, the known method for the feature importance is calculated by taking the difference between the loss function metric obtained using the original model with the feature and with the feature removed from all the trees in the model. The higher the value, the higher its importance and relevance in the prediction of the target value [41]. But SHAP is a technique that can be used to measure the impact of a feature on a single prediction value compared to baseline predictions whether the case is object-level contributions of features overall feature importance of the entire dataset [42].

The reason why the SHAP is the chosen technique for feature importance is because it calculates the impact of each feature on the model output magnitude and it shows the impact of having a certain value for a given feature in comparison to the prediction made if that feature took some baseline value [42].

8.6 Experimental Results and Discussions

This section presents the environmental settings and experimental results based on three scenarios and as an overall feature importance and cost analysis.

8.6.1 Environmental Settings

For the implementation of the experiments, an HP personal computer with Intel(R) Core i7-8850U CPU @ 1.80 GHz 1.99 GHz 16 GB RAM was used for running the experiments on the Command Prompt on Microsoft Windows 10 Pro. The imbalanced-learn and Scikit Learn Python libraries with Python 3.9 were used to run SVM-SMOTE, variance threshold feature selection, RandomForest Classifier, GuassianNB, KNeighborsClassifier, Logistic Regression, and DecisionTreeClassifier techniques. XGBoost and CatBoost libraries were used to run the classifier models. Also, Matplotlib and Seaborn libraries were used to assist in plotting the feature importance barplots for XGBoost and CatBoost models. The value 0.8 threshold was used in the variance threshold technique in feature selection. The RandomForest Classifier, GuassianNB, KNeighborsClassifier, Logistic Regression, and DecisionTreeClassifier techniques were used for comparison with proposed framework models. Finally, the category label of the IoTID2020 dataset was considered as the target variable in the experiment for the proposed framework.

8.6.2 Experimental Results

For the proposed framework, the experiment was divided into three experiments, with each aspect considered one at a time as an additional step in each experiment. The analysis is based on G-mean, precision, recall, and F1-Score. The overall accuracy was not a priority metric as the data is imbalanced and the focus is on classification; therefore accuracy is not a valid metric to conclude from.

Experiment Without Data Preprocessing In this experiment, the data was fed directly to the classification models XGBoost and CatBoost to produce the results for the specified metrics and compare them against the other classifiers. As seen in Table 8.7, it has been analyzed that the XGBoost had the best performance compared to all other classifiers as it had the highest G-mean of 0.943 meaning it gave the most true results among the total number of cases examined and in the classification of the positive cases and ensures that the lack of minority classes does not affect the quality of results produced by the classifiers. Also, XGBoost had the highest precision of predicting true positives in the majority of category labels and the recall in the prediction of actual positives classified correctly. With this said, the F1-score is relatively the highest due to the high precision and recall results for the category labels. The decision tree classifier had the second best performance with a difference of 0.01 in terms of G-mean, and it had the second highest precision and recall in the majority of category labels thus second highest F1-score when compared to all other classifiers as CatBoost performance came third best with a difference between XGBoost of 0.022. Figure 8.5 shows the metric results for the top 3 performing models.

Experiment with Feature Selection In the second experiment, the same experiment was executed with the addition of the variance threshold feature selection of 0.8 to select the best features for the classification of category labels.

Results show in Table 8.8 that XGBoost also outperforms all other classifiers as it had the highest G-mean of 0.944 and also the highest precision, recall, and F1-score in the prediction and classification of the category labels for the intrusion attacks. The decision tree classifier outperformed all other classifiers in terms of precision, recall, and F1-score in the prediction of the category labels as they were comparatively high with a G-mean difference of only 0.013. CatBoost came best third as it did have quite high prediction and classification results for the category labels in terms of precision, recall, and F1-score with a G-mean difference from XGBoost of

Table 8.7 Experiment performance results without data preprocessing

	XGBoost			*CatBoost*			*RandomForest Classifier*			*KNN*			*GaussianNB*			*LogisticRegression*			*DecisionTreeClassifier*		
	ACC	**0.976**		ACC	0.970		ACC	0.931		ACC	0.835		ACC	0.559		ACC	0.793		ACC	0.965	
	GM	**0.943**		GM	0.921		GM	0.834		GM	0.691		GM	0.609		GM	0.217		GM	0.933	
	PREC	REC	F1	PREC	REC	F1	PREC	REC	F1	PREC	REC	F1	PREC	REC	F1	PREC	REC	F1	PREC	REC	F1
DoS	**1.000**	**0.998**	**0.999**	**1.000**	0.990	0.990	**1.000**	0.995	0.998	0.998	0.983	0.99	**1.000**	0.993	0.996	0.993	0.995	0.994	**1.000**	**0.998**	**0.999**
MITM ARP Spoofing	0.863	0.824	0.843	**0.870**	0.760	0.810	0.712	0.529	0.607	0.476	0.424	0.449	0.158	**0.941**	0.271	0.500	0.008	0.017	0.810	0.786	0.797
Mirai	0.978	**0.991**	**0.986**	0.970	0.990	0.980	0.942	0.965	0.954	0.895	0.892	0.893	**0.987**	0.502	0.666	0.784	0.971	0.867	0.975	0.978	0.977
Scan	**0.980**	**0.947**	**0.963**	0.900	0.930	0.950	0.867	0.892	0.879	0.623	0.714	0.665	0.190	0.214	0.201	0.384	0.114	0.175	0.934	0.945	0.940
Normal	**1.000**	0.968	0.983	**1.000**	0.960	0.980	0.969	0.892	0.929	0.669	0.592	0.628	0.473	0.838	0.604	0.890	0.523	0.659	0.993	**0.975**	**0.984**

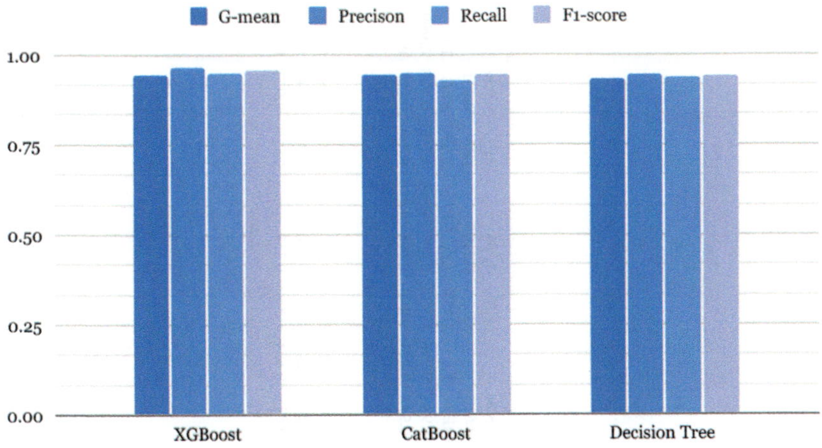

Fig. 8.5 Experiment (1): G-mean, precision, recall, and F1-score

0.025. Figure 8.6 shows the metric results for the top 3 performing models for this experiment.

Experiment with Feature Selection and Oversampling In the last experiment, the SVM-SMOTE oversampling was added. Table 8.9 displays the fact that XGBoost still outperforms all classifiers with the highest G-mean of 0.932 along with the best results for the precision, recall, and F1-score in the majority of the category labels. CatBoost classifier came second best in performance in terms of precision, recall, and F1-score with a G-mean difference of 0.008. Decision tree classifier was the third in best performance after the CatBoost classifier. Figure 8.7 shows the evaluation results for the highest three performing models.

Overall Experimental Results As a whole, the XGBoost algorithm had outperformed all other classification algorithms in the three experiments with the highest metric results for precision, recall, f1-score, and G-mean although its G-mean stayed comparatively similar with a slight increase and decrease in the experiments. Yet, it can be analyzed in the second and third experiments that the recall has been improved for the basic classifiers KNN and random forest due to the presence of oversampling.

Table 8.8 Experiment performance results with feature selection

	XGBoost			*CatBoost*			*RandomForest Classifier*			*KNN*			*GaussianNB*			*LogisticRegression*			*DecisionTreeClassifier*		
	ACC	**0.977**		ACC	0.970		ACC	0.934		ACC	0.835		ACC	0.559		ACC	0.793		ACC	0.966	
	GM	**0.944**		GM	0.919		GM	0.835		GM	0.691		GM	0.609		GM	0.216		GM	0.931	
	PREC	REC	F1	PREC	REC	F1	PREC	REC	F1	PREC	REC	F1	PREC	REC	F1	PREC	REC	F1	PREC	REC	F1
DoS	**1.000**	**0.998**	**0.999**	**1.000**	0.990	**1.000**	**1.000**	0.995	0.998	0.998	0.983	0.990	**1.000**	0.993	0.996	0.993	0.995	0.994	**1.000**	**0.998**	**0.999**
MITM ARP Spoofing	**0.891**	0.824	**0.856**	**0.870**	0.760	0.810	0.718	0.534	0.612	0.476	0.424	0.449	0.158	**0.941**	0.271	0.400	0.008	0.016	0.811	0.777	0.794
Mirai	0.977	**0.992**	**0.985**	0.970	0.990	0.980	0.944	0.970	0.957	0.895	0.892	0.893	**0.987**	0.502	0.665	0.785	0.970	0.868	0.976	0.980	0.978
Scan	**0.984**	**0.949**	**0.966**	0.970	0.930	0.950	0.880	0.890	0.885	0.623	0.714	0.665	0.190	0.214	0.201	0.372	0.106	0.165	0.938	**0.949**	0.943
Normal	**1.000**	0.968	**0.983**	**1.000**	0.960	0.980	0.976	0.884	0.928	0.669	0.592	0.628	0.473	0.838	0.604	0.860	0.552	0.673	0.989	**0.971**	**0.980**

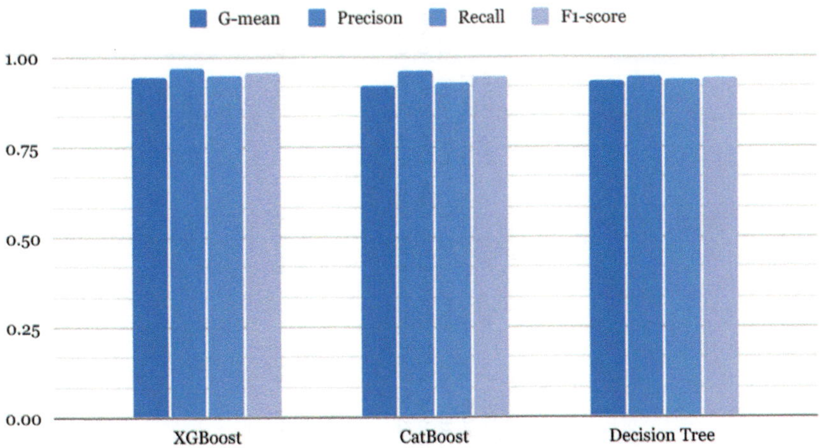

Fig. 8.6 Experiment (2): G-mean, precision, recall, and F1-score

8.6.3 *Feature Importance Results*

The effect of the classifiers XGBoost and CatBoost in different condition experiments has played a role in identifying the influential features that improve the prediction power of classification of the category type of intrusion attack. Based on each of the three experiments, each had different results for each classifier algorithm. The feature importance SHAP focused on the impact of the highest ten influential feature variables.

Experiment (1) No heavy data preprocessing was added to this experiment, and the data was fed to the classifiers directly. Based on the bar plot shown in Fig. 8.8 for XGBoost, it can be analyzed that the feature Flow_Duration showed to have the highest impact on the model output for classification of all the category types where the Dst_Port feature was the second most important variable for all category types except the DoS although it also had less prediction impact for predicting the scan class yet higher impact in predicting the normal class. The prediction impact for Mirai and MITM was similar for both features.

For the CatBoost feature importance in Fig. 8.9, it is concluded that the Src_Port feature had the highest impact on the prediction of the category type classification for category labels. Flow_Duration was the second most important variable for detection although it has less impact

Table 8.9 Experiment performance results with feature selection and oversampling

	XGBoost			*CatBoost*			*RandomForest Classifier*			*KNN*			*GaussianNB*			*LogisticRegression*			*DecisionTreeClassifier*		
	ACC	**0.966**		ACC	0.950		ACC	0.919		ACC	0.814		ACC	0.587		ACC	0.742		ACC	0.946	
	GM	**0.932**		GM	0.924		GM	0.850		GM	0.733		GM	0.625		GM	0.721		GM	0.911	
	PREC	REC	F1	PREC	REC	F1	PREC	REC	F1	PREC	REC	F1	PREC	REC	F1	PREC	REC	F1	PREC	REC	F1
DoS	**1.000**	**1.000**	**1.000**	**1.000**	0.990	**1.000**	**1.000**	0.995	0.998	0.998	0.985	0.991	**1.000**	0.966	0.982	0.993	0.995	0.994	**1.000**	**0.998**	**0.999**
MITM ARP Spoofing	**0.781**	0.794	**0.788**	**0.710**	0.780	0.750	0.575	0.580	0.577	0.387	0.525	0.446	0.162	**0.920**	0.275	0.199	0.395	0.264	0.645	0.739	0.689
Mirai	0.978	**0.980**	**0.979**	0.970	0.960	0.970	0.948	0.940	0.944	0.912	0.843	0.876	**0.980**	0.547	0.702	0.961	0.709	0.816	0.972	0.955	0.964
Scan	**0.945**	**0.939**	**0.942**	0.920	0.940	0.930	0.846	0.894	0.869	0.590	0.724	0.650	0.207	0.241	0.223	0.441	0.847	0.580	0.915	**0.949**	0.932
Normal	**0.989**	**0.960**	**0.974**	**0.980**	**0.960**	0.970	0.934	0.913	0.923	0.612	0.671	0.640	0.565	0.812	0.667	0.657	0.823	0.731	0.942	**0.939**	**0.940**

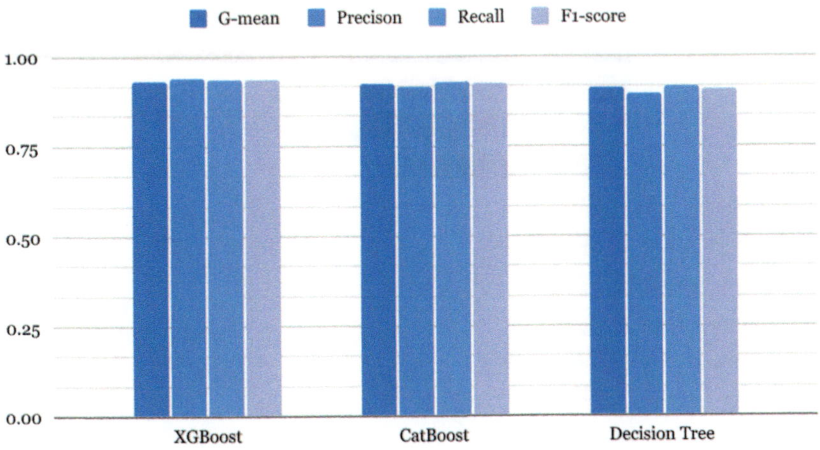

Fig. 8.7 Experiment (3): G-mean, precision, recall, and F1-score

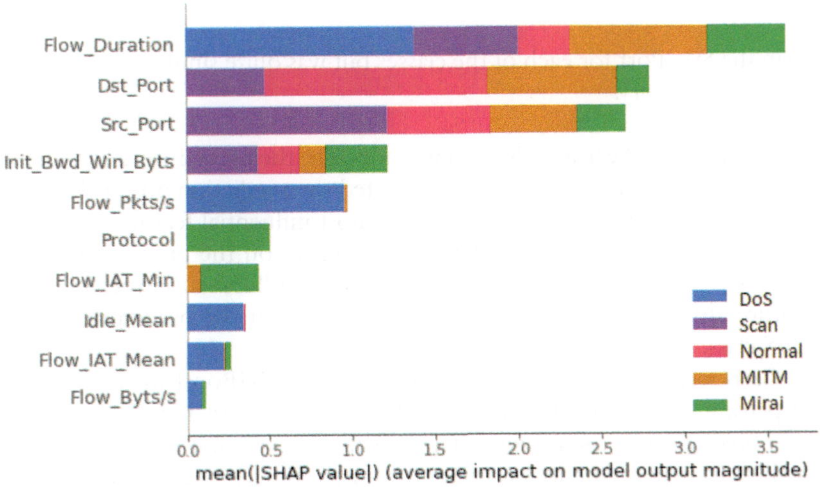

Fig. 8.8 Experiment(1): Highest ten ranked features in XGboost

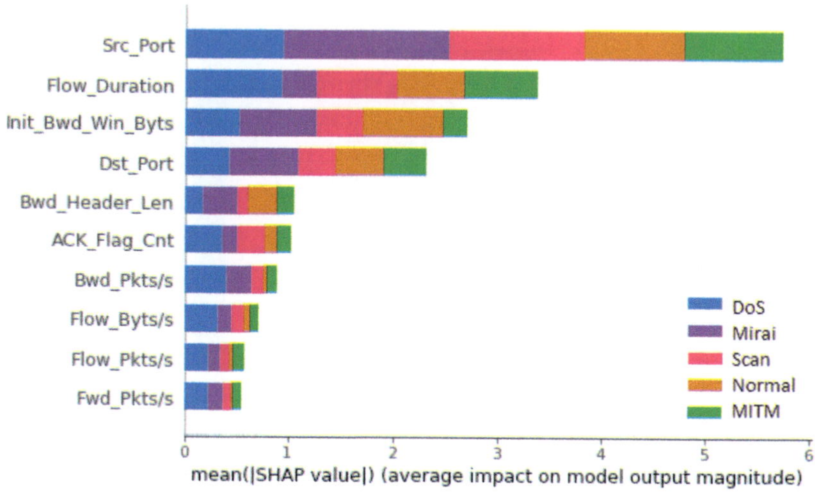

Fig. 8.9 Experiment(1): Highest ten ranked features in CatBoost

than the Src_Port for each of the classes but was quite similar in predicting the DoS category.

Experiment (2) Feature selection was added to the procedure, and the feature variables that influenced and impacted the prediction power differed. For XGBoost, Fig. 8.10 shows that the most influential feature variable is the Flow_Duration as it has the highest impact on the prediction of the category type classes, while the second most influential is the Src_Port feature although it did not have much impact on the prediction of the DoS class.

CatBoost differs slightly when compared to XGBoost as in Fig. 8.11 displaying the Src_Port feature variable as the highest impact on detection prediction of the category types, while Flow_Duration came second highest for the prediction although it can be noticed that it had a higher impact on predicting the DoS class than Src_Port. Yet, Src_Port was more impactful on the other classes in comparison.

Experiment (3) As the feature selection and oversampling techniques were added to the experiments, the change of effect on the influential features for a prediction made different impacts on classification for the

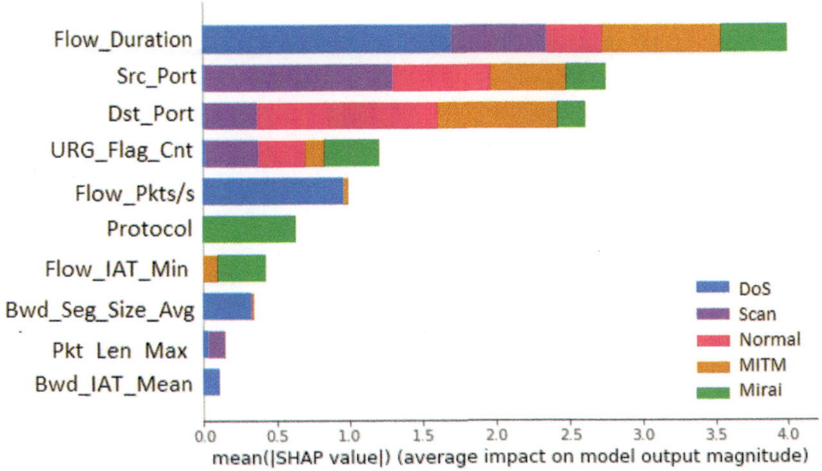

Fig. 8.10 Experiment(2): Highest ten ranked features in XGboost

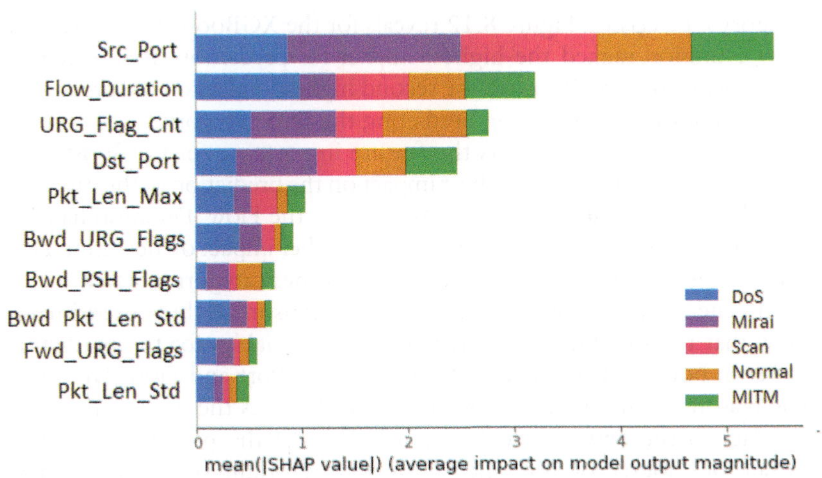

Fig. 8.11 Experiment(2): Highest ten ranked features in CatBoost

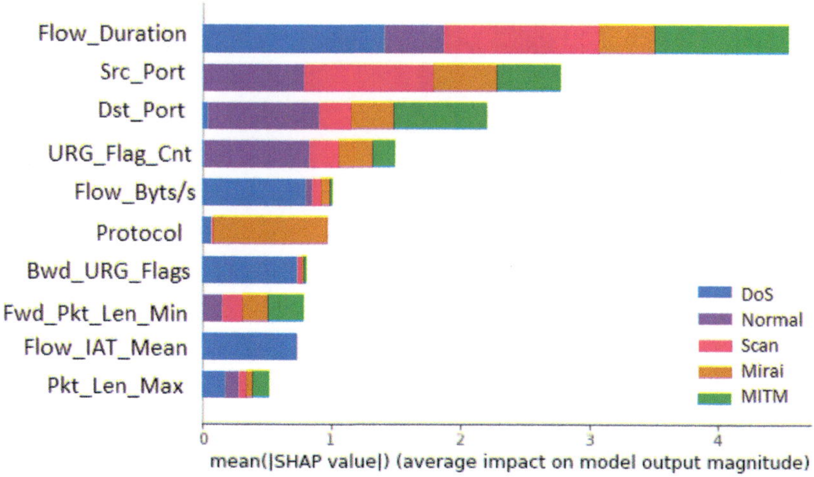

Fig. 8.12 Experiment(3): Highest ten ranked features in XGBoost

category type classes. Figure 8.12 reveals for the XGBoost that the feature Flow_Duration ranked the highest impact for prediction of all category classes, while the Src_Port had the second highest impact on the prediction although not much impact on predicting the DoS category.

Figure 8.13 that represents the feature importance for CatBoost indicates that Src_Port had the highest impact on the prediction of the category type classes classification for all types, whereas the Flow_Duration had the next highest impact and importance with a higher impact on detecting DoS in comparison, while Src_Port predicted all other categories better.

In general, feature importance interpretations in all three experiments show that the two highest influential feature variables for the prediction power in both XGBoost and CatBoost are Src_Port and Flow_Duration. The reason for this is that the source port identifies the process that sent the data to the network, so it could indicate if the packets came from a malicious source or not since it shows the origin and destination of a given flow in the network, while the flow duration feature shows the total duration of a flow in seconds indicating whether the flow pattern is suspicious or not.

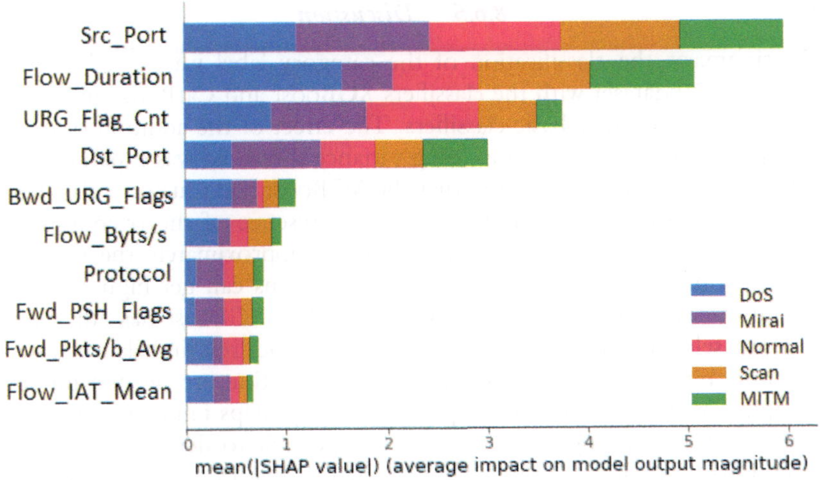

Fig. 8.13 Experiment(3): Highest ten ranked features in CatBoost

Table 8.10 Time taken for each classification model in seconds

Time taken for each model/s		
	XGBoost	*CatBoost*
Experiment (1)	*2.50*	*24.93*
Experiment (2)	*2.25*	*23.31*
Experiment (3)	*7.16*	*40.11*

8.6.4 Cost Analysis

For the detection and classification of IoT intrusion attacks from an IoT network, it requires many crucial steps along with specific resources for its process. All 625783 samples from the IoT home environment were needed for the experiments for reliable prediction. Thus, it was very time- and power-consuming due to the power and time needed for the heavy data preprocessing, training, and testing of data in classification prediction and evaluation in the procedure along with feature importance. In addition to this, the experiment was repeated three times as mentioned. Thus, the power and time needed were tripled. Based on the environmental settings, Table 8.10 shows the time taken for each classifier model separately without the rest of the experiment time taken into consideration.

8.6.5 *Discussion*

In summary, the classification of the category label was tested on the IoTID2020 dataset with the classifiers XGBoost and CatBoost and compared with the other basic classifiers. The effect of the addition of oversampling and feature selection using variance threshold was experimented in two experiments concluding that the XGBoost and CatBoost classifiers have only made a small improvement in presence of the oversampling and feature selection, yet the results stayed approximately the same in all experiments stating that these two algorithms can get high accurate results without heavy data preprocessing. Also, XGBoost has automatic feature selection; it has internal features that address imbalance distribution. Yet, the oversampling and feature selection are needed for the simple classifiers as the oversampling especially helps improve the recall. Furthermore, although all features are important to detect and prevent IoT intrusions, it is essential knowing which features have the highest influential impact on predicting the category types to get accurate results fast, and in the experiments, it was shown how the two features Src_Port and Flow_Duration play a fundamental role in prediction. As an overall, the XGBoost performed best in all conditions in experiments validating how powerful and reliable the algorithm is in predicting the category labels of the intrusion attacks.

The experiments were limited to the classification of the category labels, which could extend to the subcategory labels for the IoTID2020 dataset. Also, the SVMSMOTE oversampling was only considered and was not compared with other oversampling methods with different ratios. It also did not consider automatic clustering and data reduction although it could provide more insight toward the consumption behavior on different regions of the data distribution and undersample the data. Additionally, different feature selection techniques and techniques for predicting the most important feature variables were not taken into account. Moreover, it is noted the specific distribution of the activities for the IoTID20 dataset, and this should be tested on different datasets having a different distribution of activities for validation. Another limitation to the experiments is the lack of prior experience and repetition of the experiment as the experiments should be run repeatedly on an average of 30 times to get the mean and standard deviation for reliable results.

8.7 Conclusions and Future Works

This chapter proposes an approach for intrusion detection for the recent IoTID2020 dataset. The proposed approach includes:

- Variance threshold feature selection of 0.8 threshold applied on data
- SVM-SMOTE oversampling technique applied on feature selected data.
- Generate a multiclass classification model of the feature selected and oversampled data by the XGBoost classification algorithm.
- Generate a multiclass classification model of the feature selected and oversampled data by the CatBoost classification algorithm.
- Evaluating the models using the testing data in terms of accuracy, precision, recall, f1-score, and G-mean.
- Compare the evaluated models with other basic classifiers (logistic regression, knn, decision tree, and random forest).
- Select the most important features that influence prediction power.

The aim of the proposed approach was to develop a recent multiclass classification to classify the category type labels of IoT intrusion attacks with the application of the feature selection method variance threshold to adopt valid predictions with solving imbalanced distribution with SVM-SMOTE oversampling based on XGBoost and CatBoost algorithms to view their performance against basic classifiers in addition to specifying the influencing feature variables that play a big role in the prediction power which are Src_Port and Flow_Duration. Results show that XGBoost outperformed CatBoost along with all other algorithms on the selected IoTID2020 dataset for producing high-quality results for IoT intrusion detection and classification.

For future work and the engagement in the resource process leading to recommended actions for future improvement and research, multiple aspects can be considered such as data preprocessing in terms of automatic clustering and data reduction, along with different oversampling techniques (with different parameters) and feature selection methods. The subcategory label from the dataset can be used to detect and classify the intrusion attacks subcategory. In addition, the proposed approach can be applied and tested on different IoT intrusion attack datasets.

References

1. Ezechina M, Okwara K, Ugboaja C (2015) The internet of things (IoT): a scalable approach to connecting everything. Int J Eng Sci 4(1):09–12
2. Khan R, Khan SU, Zaheer R, Khan S (2012) Future internet: the internet of things architecture, possible applications and key challenges. In: 2012 10th International Conference on Frontiers of Information Technology, pp 257–260
3. Ashton K et al (2009) That 'internet of things' thing. RFID J 22(7):97–114
4. Evans D (2011) How the next evolution of the internet is changing everything, p 11
5. Chaudhary S, Johari R, Bhatia R, Gupta K, Bhatnagar A (2019) Craiot: concept, review and application(s) of IoT. In: 2019 4th International Conference on Internet of Things: Smart Innovation and Usages (IoT-SIU), pp 1–4
6. Reddy AN, Marks AM, Prabaharan SRS, Muthulakshmi S (2017) IoT augmented health monitoring system. In: 2017 International Conference on Nextgen Electronic Technologies: Silicon to Software (ICNETS2), pp 251–254
7. Razalli H, Alkawaz MH, Suhemi AS (2019) Smart IoT surveillance multi-camera monitoring system. In: 2019 IEEE 7th Conference on Systems, Process and Control (ICSPC), pp 167–171
8. Krasniqi X, Hajrizi E (2016) Use of IoT technology to drive the automotive industry from connected to full autonomous vehicles. IFAC-PapersOnLine 49(29):269–274
9. Kim T-H, Ramos C, Mohammed S (2017) Smart city and IoT
10. Kumar CS (2017) Correlating internet of things. Int J Manag (IJM) 8(2):68–76
11. Williams R, McMahon E, Samtani S, Patton MW, Chen H (2017) Identifying vulnerabilities of consumer internet of things (IoT) devices: a scalable approach. In: 2017 IEEE International Conference on Intelligence and Security Informatics (ISI), pp 179–181
12. Meneghello F, Calore M, Zucchetto D, Polese M, Zanella A (2019) IoT: Internet of threats? A survey of practical security vulnerabilities in real IoT devices. IEEE Internet Things J 6(5):8182–8201

13. Shojafar M, Taheri R, Pooranian Z, Javidan R, Miri A, Jararweh Y (2019) Automatic clustering of attacks in intrusion detection systems. In: 2019 IEEE/ACS 16th International Conference on Computer Systems and Applications (AICCSA), pp 1–8
14. Alaiz-Moreton H, Aveleira-Mata J, Ondicol-Garcia J, Muñoz-Castañeda AL, García I, Benavides C (2019) Multiclass classification procedure for detecting attacks on MQTT-IoT protocol. Complexity 2019:1–11
15. Dash PB, Nayak J, Naik B, Oram E, Islam SH (2020) Model based IoT security framework using multiclass adaptive boosting with smote. Secur Privacy 3(5):e112
16. Tarekegn AN, Giacobini M, Michalak K (2021) A review of methods for imbalanced multi-label classification. Pattern Recogn 118:107965
17. Qaddoura R, Al-Zoubi AM, Faris H, Almomani I (2021) A multi-layer classification approach for intrusion detection in IoT networks based on deep learning. Sensors 21(9):2987
18. He H, Garcia EA (2009) Learning from imbalanced data. IEEE Trans Knowl Data Eng 21(9):1263–1284
19. Weiss GM, McCarthy K, Zabar B (2007) Cost-sensitive learning vs. sampling: which is best for handling unbalanced classes with unequal error costs? Dmin 7(35–41):24
20. Abdi L, Hashemi S (2015) To combat multi-class imbalanced problems by means of over-sampling techniques. IEEE Trans Knowl Data Eng 28(1):238–251
21. Wang S, Yao X (2012) Multiclass imbalance problems: analysis and potential solutions. IEEE Trans Syst Man Cybern Part B (Cybernetics) 42(4):1119–1130
22. Sharafaldin I, Lashkari AH, Ghorbani AA (2018) Toward generating a new intrusion detection dataset and intrusion traffic characterization. In: ICISSp, pp 108–116
23. Qaddoura R, Al-Zoubi A, Almomani I, Faris H (2021) A multi-stage classification approach for IoT intrusion detection based on clustering with oversampling. Appl Sci 11(7):3022
24. Yang X, Kuang Q, Zhang W, Zhang G (2018) AMDO: an over-sampling technique for multi-class imbalanced problems. IEEE Trans Knowl Data Eng 30(9):1672–1685

25. Panigrahi R, Borah S, Bhoi AK, Ijaz MF, Pramanik M, Kumar Y, Jhaveri RH (2021) A consolidated decision tree-based intrusion detection system for binary and multiclass imbalanced datasets. Mathematics 9(7):751
26. Iwendi C, Khan S, Anajemba JH, Mittal M, Alenezi M, Alazab M (2020) The use of ensemble models for multiple class and binary class classification for improving intrusion detection systems. Sensors 20(9):2559
27. Miah MO, Shahriar Khan S, Shatabda S, Farid DM (2019) Improving detection accuracy for imbalanced network intrusion classification using cluster-based under-sampling with random forests. In: 2019 1st International Conference on Advances in Science, Engineering and Robotics Technology (ICASERT), pp 1–5
28. Zheng D, Hong Z, Wang N, Chen P (2020) An improved LDA-based ELM classification for intrusion detection algorithm in IoT application. Sensors 20(6):1706
29. Elrawy MF, Awad AI, Hamed HF (2018) Intrusion detection systems for IoT-based smart environments: a survey. J Cloud Comput 7(1):1–20
30. Chen T, Guestrin C (2016) Xgboost: a scalable tree boosting system. In: Proceedings of the 22nd ACM SIGKDD International Conference on Knowledge Discovery and Data Mining, KDD'16. Association for Computing Machinery, New York, pp 785–794 [Online]. Available: https://doi.org/10.1145/2939672.2939785
31. Chen T, Guestrin C (2016) Xgboost: a scalable tree boosting system. In: Proceedings of the 22nd ACM SIGKDD International Conference on Knowledge Discovery and Data Mining, pp 785–794
32. Chen M, Liu Q, Chen S, Liu Y, Zhang C-H, Liu R (2019) XGBOOST-based algorithm interpretation and application on post-fault transient stability status prediction of power system. IEEE Access 7:713149–13158
33. Chen T, He T, Benesty M, Khotilovich V, Tang Y, Cho H et al (2015) XGBoost: extreme gradient boosting. R package version 0.4-2 1(4):1–4

34. Huang G, Wu L, Ma X, Zhang W, Fan J, Yu X, Zeng W, Zhou H (2019) Evaluation of CatBoost method for prediction of reference evapotranspiration in humid regions. J Hydrol 574:1029–1041 [Online]. Available: https://www.sciencedirect.com/science/article/pii/S0022169419304251
35. Zyphur MJ, Pierides DC (2017) Is quantitative research ethical? Tools for ethically practicing, evaluating, and using quantitative research. J Bus Ethics 143(1):1–16
36. Queirós A, Faria D, Almeida F (2017) Strengths and limitations of qualitative and quantitative research methods. Eur J Educ Stud 3:369–387
37. Ullah I, Mahmoud QH (2020) A scheme for generating a dataset for anomalous activity detection in IoT networks. In: Canadian Conference on Artificial Intelligence. Springer, pp 508–520
38. Corchado Rodríguez J (2013) Advances in Artificial Intelligence
39. García S, Luengo J, Herrera F (2015) Data preprocessing in data mining. Springer, Cham, Switzerland vol 72
40. Obiedat R, Qaddoura R, Ala'M A-Z, Al-Qaisi L, Harfoushi O, Alrefai M, Faris H (2022) Sentiment analysis of customers' reviews using a hybrid evolutionary SVM-based approach in an imbalanced data distribution. IEEE Access 10:22260–22273
41. Meidan Y, Bohadana M, Shabtai A, Ochoa M, Tippenhauer NO, Guarnizo JD, Elovici Y (2017) Detection of unauthorized IoT devices using machine learning techniques, arXiv preprint arXiv:1709.04647
42. Parsa AB, Movahedi A, Taghipour H, Derrible S, Mohammadian AK (2020) Toward safer highways, application of XGBoost and SHAP for real-time accident detection and feature analysis. Accident Anal Prevent 136:105405

CHAPTER 9

Malware Mitigation in Cloud Computing Architecture

Sai Kumar Medaram and Leandros Maglaras

9.1 Introduction

Cloud computing is one of the decade's most trending discussions in information technology (IT). A preponderant of IT either has integrated or has plans to adopt products and services around the cloud computing paradigm. Cloud computing is defined as a model for providing on-demand, convenient and ubiquitous network access to a shared pool of computing resources that can be configured (such as storage, networks, servers, services and applications) and may be provisioned rapidly and released with little interaction with the service provider or little management effort. "Cloud" itself is a shared resource which is widely influential since it is not merely shared among a large volume of users but offers

S. K. Medaram
CTI, De Montfort University, Leicester, UK
e-mail: Sai.Kumar@dmu.ac.uk

L. Maglaras (✉)
School of Computing, Edinburgh Napier University, Edinburgh, Scotland
e-mail: l.maglaras@napier.ac.uk

I. Almomani et al. (eds.), *Cyber Malware*, Security Informatics and Law Enforcement, https://doi.org/10.1007/978-3-031-34969-0_9

dynamic access which is dependent on the demands (Ou, 2015). The cloud is not an array of software, hardware or services. It is an integration and combination of vast provisions in information technologies. In the cloud environment, users do not need to possess the infrastructure enabling various computing services. But much more, the services become accessible to a computer from any location in the world. The features integrated into the environment include those offering support multi-tenancy and high scalability and enhanced flexibility compared to older methodologies for computing. It can be used to allocate, reallocate or deploy resources dynamically while being able to monitor their performance continuously. Cloud computing has four deployment models (public, private, hybrid and community), and three service models (infrastructure as a service, IaaS; software as a service, SaaS; and platform as a service, PaaS), which provides a description of the relationship that exists between cloud service producers and cloud service consumers. Thus, a user can access one or multiple cloud deployment models. However, the increased adoption of cloud services and products has met a growth of malicious activities, codes and programs targeted at the infrastructure. Even though the potentials of cloud computing are yet to be fully tapped, public consent already reveals security as its most critical flaw at the moment. Many of these activities and attacks are generically described as security threats that dissuade users from exploring these benefits. Nowadays, the number and severity of cyber-related attacks are on a drastic increase. Commonly reported security threats in cloud computing (CC) infrastructure include data loss and breaches, malicious insiders, account or service hijacking, identity theft, phishing attacks, man-in-the-middle attacks, denial of service (DOS), distributed denial of service (DDOS) attacks, cookie poisoning attacks, wrapping attacks, etc. [1]. In general, several variants of malware are the reason for these attacks. Malware is any type of software which put harmful and malicious effects on the OS (operating system), software or other components. It is designed with the intention to cause harm or damage to its target system. Trojan horses, worms, backdoors, viruses, spyware, rootkits, ransomware and botnet are typical examples of malware [2]. Each variant and family of the malicious code is designed for peculiar purposes. While some variants of malware steal sensitive data, many others initiate DDoS attacks and give room for remote code execution [3]. Highly sophisticated attacks employ more than one type and family of malware. The amount of malware samples has increased rapidly over the years.

According to scientific and business reports, about one million variants of malicious software are generated on a daily basis [2]. The majority of these surfacing variants of malware are evolving/modified versions of malware previously in existence. The number of malware-related attacks has increased tangibly due to the addition of new devices, e.g. IoT devices to the

computer networks daily, the volume of data generated daily on social as well as the number of applications built in a compressed period of time. The complexity of malware attacks, strategies for spreading as well as economic damage to economies across the globe have recently hit a peak. Research has it that these attacks cause damage in trillion dollars to the world economies [2]. Malware detection is, therefore, the procedure for specifying whether a particular program or code is benign or malware. The massive and continuously growing ecosystem of malicious tools and software constitutes a daunting challenge for IT administrators and network operators. Various methods exist for the detection of malware and may be broadly categorised as traditional and new approaches. Traditional methods include heuristic-, behaviour-, signature- and model checking-based, while the new methods include mobile device-based and deep-learning detection [4–6]. Some of these existing techniques are precise with detecting specific kinds of malware while being unable to identify other types, or even new variants of the same type. For example, the signature-based method works optimally with known and various versions of the same malware, while it fails in detecting unknown malware that possesses a totally different signature. Heuristic-, behaviour- and model checking-based methods of detection can help in detecting a reasonable part of the zero-day malware. Unfortunately, they are unable to detect new malware that employs advanced packing techniques [2].

It was posited over a decade ago that we can expect to experience several security exploitations with cloud service providers as well as users, which will shift research focus to fixing these loopholes. Hence, we are experiencing a drastic evolution in the CC discipline with underlying efforts to address the security and privacy issues raised by this paradigm. Therefore, research towards detecting these malware as well as safeguarding the cloud architecture against malware attacks are increasing. This necessitates this research which seeks to analyse malware mitigation strategies in cloud computing architecture.

The contributions of this chapter are:

- systematically analyse the security and malware threats in cloud computing architecture.

- examine malware detection methods in cloud computing infrastructures.
- examine the techniques for safeguarding against malware challenges in cloud computing infrastructure.
- make recommendations on the applicability of these techniques.

This research work will be beneficial for both corporate/institutional and individual parties who provide or utilise cloud services, as well as those considering adopting cloud computing provisions. It will also help service providers, engineers and professionals to be abreast with malware challenges in cloud computing as well as the techniques for safeguarding this architecture against malware challenges. Cloud security service providers would find immense treasure in this research. The recommendations that will follow the findings will help facilitate the activities of these security providers, security analysts and threat intelligence professionals.

This chapter comprises five sections. Section 9.1 (introduction) contains an overview of the study, as well as its aim and objectives, the expected impacts of the project and a brief of the whole project's structure. Section 9.2 contains the examination of the relevant literature, especially on security threats in cloud computing infrastructure. Section 9.3 contains the malware detection methods in cloud computing infrastructure. Section 9.4 contains the techniques for safeguarding against the malware challenges in cloud computing. Section 9.5 contains discussion and analysis, while Sect. 9.6 comprise the conclusion and recommendations.

9.2 Cloud Computing Structure and Deployment

"Cloud" is a shared resource which is widely influential since it is not merely shared among a large volume of users but offers dynamic access which is dependent on the demands (Ou, 2015). The term "Cloud" stems from the fact that there is an abstract boundary, dynamic change of scale and ambiguous location, which mimics an actual natural cloud (Ou, 2015), even though there is no such existence in the actual world. The cloud is not an array of software, hardware or services. It is an integration and combination of vast provisions in information technologies. Furthermore, because of the continuous introduction of new technologies to the cloud, the cloud size keeps increasing. Apart from this, the US Department of Commerce describes "Cloud computing as a model for providing on-demand, convenient and ubiquitous network access to a shared pool of computing resources that can be configured (such as storage, networks, servers, services and applications) may be provisioned rapidly and released with little interaction with the service provider or little management effort. Cloud Computing is simply the combination of a platform technology that offers storage and hosting and service. It is comprised of five major features, four deployment models and three service models". Users in this environment do not need to possess their peculiar infrastructure to carry out various computing services. In fact, any user can have access to these from any computer from any location in the world. This integrates characteristics facilitating high multitenancy and scalability and providing increased flexibility when compared to conventional computing methodologies. It can be used to allocate or reallocate and deploy resources dynamically with the provision for continuous monitoring of their performance. Although the definitions of cloud computing (CC) have a lot of variations, certain fundamental principles underline this trending computing paradigm. CC has provisions for technological capabilities – for off-premises maintenance – which are to be delivered on-demand through the Internet [7]. Users of resources in the cloud do not own the resources but only pay for the resources on a pro-rata basis since the party public cloud services and resources are owned and managed by a third party, therefore virtualisation of a principal concept. Also, the cloud services and architecture comprise certain functionalities which are key in defining the features of cloud architecture. According to an article by NIST's Cloud Computing Standards Roadmap [8], which states typical features of a cloud infrastructure to include:

(i) self-service-on-demand: an individualistic and automatic offering of computing capabilities (server time and network storage) while eliminating the intervention of human agency.
(ii) Broad network access, cloud offering capabilities and features available across the network by the use of tools which facilitate use by diverse clients (e.g. desktops, tabs, mobile devices).
(iii) Resource pooling. Cloud computing resources are open to the service of multiple users on a need basis. These various virtual and physical resources (storage, processing, memory, and network bandwidth) are assigned actively according to the demands of the consumer.

(iv) Location independence: The location of the service is virtually out of the authority of the customer. The consumer may only have an idea of possible locations of the infrastructure.
(v) Rapid elasticity. The capabilities and services provided and released are automatic and elastic in some cases. Also, based on the demand for resources, the infrastructure should be able to scale rapidly and be appropriated dynamically.
(vi) Measured service. The cloud service ensured optimised resource utilisation through leverage of measuring capability of service type at some level.

Generally, CC combines traditional networking technologies and computing methods, e.g. parallel computing, distributed computing, network storage technologies, utility computing, virtualisation, high available, load balance, etc. [9]. For example, distributed computing is aimed at breaking down large computations into little segments and allocating multiple computers to do the calculation, collection and assembling of all results (Ou, 2015). Meanwhile, parallel computing brings together a large volume of computational resources in order to process a particular task, which constitutes an effective solution to parallel problems [10]. Moreover, technologies for network-attached storage (NAS) link storage devices with a set of computers through the standard network topology. This network-attached storage meets the requirement for rapidly increasing storage volumes and offering adequate space for storage for the connected hosts. Meanwhile, storage area network (SAN) is another technology for network storage, which uses a Fibre Channel to link a group of computers without standard topology, which is often used in an environment with high-volume storage. However, the technologies mentioned above are just part of cloud computing, which indirectly indicates the massive scale of CC. Until now, several prominent information technology firms have used and deployed CC development because of its potential for revolutionary technology and commercial value.

9.2.1 Historical Perspective

The concept "Cloud" can be traced to the 1950s; at a time, the mainframe computer was gaining acceptance in the area of computation, deemed to be the future of computing, and was becoming attractive in corporations as well as academia. Nevertheless, as a result of the lack of capacities for internal processing as well as access by client computers, a proposal was put forward to support the idea that many users be allowed physical sharing and access to the computer as well as CPU time from more than one (many) terminals. It was popularly regarded across the industry as time-sharing [11]. Therefore, the "Cloud" rudiment was built. Sun Microsystems, in 1983, brought in the concept of "the network is the Computer", which transcends the conventional boundary of the computer. Amazon, in 2006, released its Elastic Computer Cloud service, which enables resizable computing capacity, which developers were excited about because the web-scale cloud computing was optimised. Apart from this, the computing resources can be totally controlled with a high capacity for scalability with an adjustment of computing demands. On August 9, 2006, Google CEO, Eric Schmidt, brought in the idea of "cloud computing", which was premised on "Google 101", a project by Christophe Bisciglia. In 2007, IBM and Google began to promote CC in universities across the United States, such as Massachusetts Institute of Technology; Carnegie Mellon University; Maryland University; University of California, Berkeley; and Stanford University. The project was targeted at lessening distributed computing's cost in academic investigations and also offered support for software, hardware and technique. In 2008, Google, in collaboration with National Chiao Tung University and National Taiwan University, launched "The Cloud Computing Project" in Taiwan for a massive campaign of this technology on campus. In 2008, IBM launched its' first-ever centre for cloud computing in Wuxi, China. On July 29, 2008, Intel, HP and Yahoo disclosed an associated research program in Singapore, Germany and the United States, which was targeted at establishing six data centres, with each data centre designed with 1400 to 4000 processors. On the same 2008, on August 3, Dell was enforcing the trademark right, which was targeted at reinvigorating the fact that the control power may remodel the technical architecture. Still, in 2008, Microsoft introduced Microsoft Azure which is an infrastructure and platform for cloud computing, providing services and applications establishment, managing and implementation through the Microsoft data centre [12]. In 2010, Rackspace, NASA, Dell, Intel and AMD announced an open-source project, described as "OpenStack", with the control for a large pool of computer, networking and storage resources across a data centre, which is used in the building of public and private clouds. Not long after, Oracle and IBM declared the "Oracle Cloud" and "IBM Smart Cloud" in 2012 and 2011, respectively. Judging from the above evolution and history, it is clear that cloud technology developed rapidly after 2000, and its embrace and application are increasing till today.

9.2.2 *Advantages of Cloud Computing*

Cloud computing is a technology experiencing rapid growth and is delivering attractive and amazing measurable services offering enterprises the opportunity to monetise their business and grow their productivity and profit level while simultaneously saving costs. It is keeping up to pace with the delivery of virtual, secure and economically viable solutions [1]. According to [13], cloud computing present a plethora of benefits and advantages; some of which include the following: scalability, the cloud allows enterprises to bring in computing resources whenever they are needed; masked complexity, without the users' awareness [14] and participation, maintenance and upgrades of the service or product can be carried out; ecosystem connectivity, the cloud promotes external collaboration between partners and consumers which brings about increased innovation and improvements in productivity; cost flexibility, with cloud computing, the requirement to pay license fees for software, or to finance the installation of software or building of hardware, is eliminated; and adaptability, cloud computing affords enterprises the opportunity to adjust and accommodate several user groups that comprise several assortments of devices.

In conventional computing, there is a requirement to duplicate the lessons learnt in one environment in the other. However, in CC, the improvement of some parts covers all users [13]. In CC, there is a provision to automatically scale up and down resources, whereas, in traditional computing, there is a need for the intervention of humans in order to add software and hardware. CC environments are commonly virtualised; meanwhile, traditional environments are primarily physical [13].

CC is changing the platform for service delivery and consumption, even the approach with which users and businesses interact with IT resources. The interest in the topic of cloud computing is growing across the industry. Transaction on Services Computing of IEEE, in 2008, adopted that CC be introduced into the taxonomy as an area in computing services [13]. The European Commission 2012 outlined a computing strategy to facilitate the drastic embrace of CC in every sector of the economy. Because several pieces of research in CC were funded, e.g. ARTIST, CLOUDMIG,

REMICS, etc. [13], many companies since then have been migrating to the CC model, while others are evaluating their transition. In collaboration with Economist Intelligence Unit 2011, IBM carried out a survey which engaged 572 business as well as technology executives from all over the world to identify how establishments utilise CC today and what their plans for the future are [13]. Nearly 75% of establishments had adopted, piloted or substantially implemented CC in their operations (while the remaining proposed their adoption within 3 years). This survey also demonstrated that the adoption of cloud is not exclusive to big companies, as 67% of organisations with revenues lower than US1$ billion and 76% of companies with revenues between 1 and 20 billion US dollars have at some point adopted cloud computing. As it relates to quality attributes, over 31% of executives replied that cost flexibility was a strong justification for

their adoption of cloud computing. Following cost flexibility are security, scalability, masked complexity and adaptability [13]. According to Pathak et al., in [15], the primary goal of CC is to enable inexpensive and scalable on-demand computing infrastructure that presents high service levels.

9.2.3 Classification of Cloud Computing Architecture

CC may be classified on the basis of deployment models and services offered. Based on the service models, they can be categorised into three, namely, (i) software as a service (SaaS), platform as a service (PaaS) and infrastructure as a service (IaaS) [16]. IaaS is the lowest layer with the offering of service for basic infrastructural support. Meanwhile, PaaS is the middle layer, with the provision of services that are platform-oriented, apart from the provision of the environment for user applications' hosting. SaaS is the top layer with the provision of a complete application that is made available as a service on demand. The following services are enumerated [8].

(a) **Software as a Service (SaaS)** SaaS are apps that operate within the infrastructure that is provided to be used by the service user. These apps are provided from the IT resources of different clients through a thick client or web browser. Software as a Service ensures that complete apps are hosted on the web and that consumers utilise them. Payment is done on a pay-per-use approach. This eradicates the demand to install and execute the app on the local computer of the customer, thereby taking off the burden for software maintenance from the customer. In software as a service, there is the convergence coherence mechanism and the divided cloud by which all data items have either the "Write Lock" or "Read Lock" [17]. Two kinds of servers are adopted by SaaS: the domain consistence server (DCS) and the main consistence server (MCS). Cache coherence is actualised by the agreement between DCS and MCS [18]. In this infrastructure, if the main consistence server is compromised or damaged, there is a consequential loss of control over the cloud environment. Therefore, the security of the MCS is a vital requirement.

(b) **Platform as a Service (PaaS)** This enables the user of the service to deploy apps on the cloud infrastructure, apps built using libraries, tools, languages and services of the service provider. The provision also comprises an environment for software execution. For instance, there can be a Platform as a Service app server which affords the lone developers to deploy applications based on the web without the need to buy actual servers and carry out set-up. This model targets the protection of data, which is very paramount, especially in storage as a service. In the event of congestion, there can be the challenge of cloud environment outage. Therefore, the requirement of security to prevent outages is vital in ensuring load-balanced service. For security reasons, the data is required to be encrypted whenever it is hosted on a platform. There has been the proposition of CC architectures that employ multiple techniques for cryptography in order to provide cryptographic cloud storage.

(c) **Infrastructure as a Service (IaaS)** This is concerned with hardware resources' sharing for services execution, typically by the use of virtualisation technology. Potentially, by the use of IaaS, several consumers utilise available resources. These resources may be scaled up easily depending on the user's demand, and payments are ideally on a pay-per-use basis. These all require management since they are virtual machines. Therefore, there is a requirement for a governance framework to regulate the creation as well as the usage of virtual machines. This helps to also prevent unsanctioned access to sensitive information of users [19]. This is a provision that affords access to the platform to give room for the consumer to access services of networks, storage, processing, etc., to enable the consumer to access applications and operating systems that necessitate service provision through the provision of the infrastructure [20].

In the enumerated services model, cloud infrastructure's management transcends the control of the service user. For example, in SaaS (application configuration settings, network, operating systems, storage, servers, etc.), in PaaS (network, storage, servers, operating systems, with the exemption of deployed apps) many components are beyond the management and control of the user, while in IaaS, operating systems, applications and storage do not have adequate control over the network's components, which lies within the consumer's control.

According to [1], the primary deployment models of the cloud computing architecture can be described as represented below:

(a) **Private Cloud** This infrastructure is found on a private network and is regularly managed by the cloud provider or internal enterprise data centre of the organisation and may reside either off-premises or on-premises. The private cloud is more secure since only the organisation that owns the resources has access to the operation and control of environments for service delivery. It is targeted at addressing the concerns or bothers in data security and provides higher control, but does not offer advantages such as reducing operational and capital costs.
(b) **Public Cloud** This is designed for a range of groups or the public, and it is owned and controlled by a cloud provider. The resources hosted on this infrastructure are dynamically provided on a pay-per-use and on-demand approach. It is, however, more vulnerable to malicious attacks and constitutes the reason for its not being so secure. It offers several advantages to its users, such as location independence, flexibility, scalability and no initial capital investment [1].
(c) **Hybrid Cloud** This particular infrastructure combines different clouds that are linked by standardised technology to share applications and data irrespective of location and ownership. It provides greater control and more flexibility over the application by combining the benefits of each other and as well addressing their limitations [21].
(d) **Community Cloud** This cloud infrastructure is made for use by several establishments that have common interests in a single community. Every participant in the community cloud has access, freely, to applications and data. Several other models of cloud deployment are being built due to the varied needs of different users. For instance, the virtual private cloud is a means of using the public cloud in a private way and using a virtual private network (VPN) to inter-connecting the resources.

9.2.4 *Areas of Application of Cloud Computing*

The industrial systems of today are typified by a strong reliance on comprehensive IT infrastructure at the site of the customer [22]. During the entire system's lifecycle, the costs of IT infrastructure, hardware and cost of maintenance are quite high. CC offers a new means of providing services and delivering industrial software to customers on demand. There are main opportunities for companies as it relates to the provision of cloud services, which turns out to heighten the competitiveness by the offering of cutting-edge cloud solutions which is to be utilised in controlling and interacting with complex industrial systems [13]. A few examples of CC in the application space include:

- Web-based email or online email: any email client accessed through the Internet and implemented as a web app. Examples include Google, Yahoo and Microsoft mails).
- Online storage services: These provide services for storing e-data using third-party services that can be accessed through the Internet (e.g. Microsoft's SkyDrive, ZumoDrive and Humyo)
- Online collaboration tools: this refers to software, social and Web tools that are used to promote website customer communication in order to attract more sales and real-time satisfaction across the Internet. These include Stixy, Google Wave, Mikogo and Spicebird
- Online office suite: this refers to a set of programs that are implemented as web-based apps used to automate conventional office tasks, such as Microsoft Office Live, ThinkFree, Ajax13 and Google Drive).

9.2.5 *Security Expectations in Cloud Computing*

Also, according to [23], the security expectations can be discussed as described below:

(a) **Data Security** Fittingly verifying data from the outside world is very vital to assure that data is ensured and has a low propensity for damage. With the uptrend in cloud computing, several vulnerabilities may surface when the information is shared in an unprecise manner within the fluctuated frameworks in CC [23]. Guaranteeing the privacy and security of data in CC implies the ability to assure the standard key security facets, namely, accessibility, integrity and confidentiality. The most vital prerequisites for the security of data are data integrity which alludes to the guaranteeing that the data of clients are not changed outside their consent or approval. In order to guarantee data integrity, from the perspectives of both the supporter and supplier, secure encryption algorithms are most often adopted. Nevertheless, mere encryption does not absolutely guarantee noxious alteration of data [24]. As a result of the circulated as well as dynamic shared nature of the cloud, privacy is another fundamental requirement for cloud clients. This alludes to exactness and data security which gives room for ensuring delicate and private information is kept so. This means that the framework of the cloud framework is expected to be made available to approved, validated clients anywhere, at any time and across any platform. There are some threats in cybersecurity that cloud service availability may be faced with and are majorly network-based attacks, e.g. DDoS attacks [25]. Meanwhile, cloud suppliers should maintain a befitting activity plan in order to handle these threats and dangers.

(b) **Cloud Network Infrastructure Security** A provider of cloud service must accept trustful network traffic and have provisions for blocking malicious ones [23]. The security infrastructure of the cloud network should be able to identify and prevent intrusions, deny and protect against DoS attacks, to enable notification and logging. Denial of service defences is anchored on network security, which must efficiently filter queries and recognise attackers in order to prevent harmful attacks [23]. The intrusion prevention and detection systems IPS and IDS, respectively, block or detect malware attacks, spam signatures and virus signatures, but some also report positive results. Moreover, logging and notification create the avenue for cloud users to have certain hints into the cybersecurity health of the network.

(c) **Cloud Applications Security** Companies are expected to protect their cloud-based apps against a vast array of cybersecurity attacks and threats. Additionally, the security of cloud apps resembles the security of web applications when they are hosted in server centres. Several businesses put out a single sign-on (SSO) which is to allow clients to have access to different individual cloud administrations [26]. In an overview, it is hard to accurately update SSO arrangements since it is anchored on a safe programming layer, which is a requirement for different confirmation strategies. The International Standards Organization gave a definition of information security as concerns or bothers, which may also be guided as it relates to the CC principal security requirements for a secure and effective technology solution. According to [1], these primary requirements include the following: (i) confidentiality, this implies keeping the data of the users and only granting access to privileged entities, (ii) integrity implies the assuring of no modification or alteration in data while being transported or stored, and access is granted to only the authorised modification, change, delete or copy data; (iii) availability implies the assurance that the services needed or the data in request by the user are steadily accessible anywhere and at any time [27]; (iv) authentication implies the verification of the user's identity before access is granted to data, which may be carried out by utilising certain protections to their profiles; and v) authorisation implies the assurance that the user that made a request to the particular data has the right of access to it [28].

9.2.6 Security Threats in Cloud Computing

Despite the massive adoption of CC so far, there still exist some aspects of these cloud computing that make several organisations not confident and excited about migrating to the cloud. As some of the characteristics of CC enable attacks and malware that is novel in nature. Also, it has been stated that malware developers has made the cloud their major target. Furthermore, certain architectural loopholes made CC vulnerable to several privacy and security threats. Despite the numerous advantages CC has introduced by its service-oriented-multi-tenancy approach, it has also opened up a worm of effect as well as issues to security and privacy of user information, as well as in seeking to improve the efficiency of asset protection [29]. For instance, the virtualisation of the software layer could lead to a vulnerability of shared physical resources within the regulation of the server, which also includes virtual machines (VMs), data and memory. In CC, the service user does not manage or control the underlying cloud infrastructure. The service provider would normally have the authority of the infrastructure. Therefore, several security risks exist in the usage of the cloud infrastructure, some at the provider and others at the consumer [29].

Generally, the threats that CC platforms encounter are very similar to those of other computing platforms. Suryateja (2020) recognised several major threats that cloud computing environments are exposed to, such as unethical and abused usage of cloud resources, shared technology vulnerabilities, application programming interface (API) vulnerability, malicious insiders, accounts, data leakage/loss service, unknown risk profile and traffic hijacking. Although several vulnerabilities and risks exist, the threats listed are popular in the CC environment; the environment also has risk factors. Tang in [16] reported these risks: inherent platform, virtualisation, storage data sharing, human resources management, security, operational management, misuse, network security, interoperability, multi-directional audit and multiparty audit. Security risks and threats are major sources of concern in cloud computing for many organisations, largely because of the physical infrastructure location dispersal as well as the data's residency, which is geographically spread. The laws for data protection are general dependent on country; therefore, data location is an issue: where data residents in a country without adequate laws to protect sensitive data, therefore making user data vulnerable [30].

9.2.6.1 Malware Threats

In the last couple of years, society is becoming increasingly reliant on technology. Computers and various devices are used for receiving emails, news and online shipping. These systems' availability and integrity are to be protected deliberately against threats [31]. Terrorists, rival corporations, amateur hackers and even foreign governments launch complex attacks against these systems. Crucial to these attacks is the use of malicious codes or malware. The word "malware" is gotten from two words, "malicious and software". This makes malware software which put harmful and malicious effects on the OS (operating system), software or other components. According to [31], various classes of malware exist, which include ordinary malware attacks and network-based malware attacks. Malware such as spyware is used to cause damage to the user's machine, in network malware, while in ordinary malware attacks, malware such as inf system.inf or like autorun is used to constitute harm to the user. Various kinds of malware exist in the cloud, namely, Trojan horses, worms, backdoors, viruses, spyware, rootkits, ransomware and botnet, plus several other kinds of software with undesirable behaviour. These malwares used various approaches to afflict the user machine. They cause damages that may range from modification of files to a denial of service or a complete shutdown of the system or service [32]. Although there are several efforts towards the prevention and detection of this malware in cloud infrastructure, the results have not been absolutely effective, even as these attacks are metamorphosing and modifying their codes, allowing them to constitute another malware that the system will be naïve towards. This is beckoning on increased research into how this malware detection and prevention or resolution in cloud computing may be enhanced.

9.2.7 *Security Attacks in CC Architecture (Including Mitigation Strategies)*

In this section, some malicious attacks on CC infrastructure, with suggested mitigation strategies are presented;

(a) **Cross-Site Scripting (XSS) Attacks** – The eavesdropper inserts malicious code into the web page of the user in order to redirect him to the website of the attacker and thereby secure access to the sensitive data of the user. According to [1], there are two ways this is being carried out, which include either the use of stored XSS (which stores malicious code permanently into a resource that is managed by the website app) or reflected XSS (which does not permanently store the malicious code, but instantly reflect it back to the user) [33]. The technique of sanitisation or content filtering employs filter functionalities to eliminate malicious data from the data of the user. These filter abilities are activated for operation after the web application has read the user input. Nevertheless, following content filtering, it is hard for some applications to remove untrusted content scripts whenever there is the allowance for HTML markup in user input. By the employment of advanced content filtering, the web browser parser of the users can anticipate untrusted content. However, there have been the discussion of some other approaches that may be employed to analyse and expose vulnerabilities in web applications [1].

(b) **Structured Query Language (SQL) Injection Attacks** – In the standard SQL code, the malicious agent uses malicious code to secure unsanctioned access to the database in order to obtain the user's sensitive data. Here, the web gives way for the SQL Server to access the hacker's data by perceiving it to be the user's data, which helps the attacker to have information about the functioning of the website and then be able to effect changes therein. Owing to the insufficiency of structural knowledge of queries generated, it has been hard to implement the various measures proposed for the filter or validation of user input. In the instances where the source code of applications is available, static analysis may be employed to validate the user input before integration into the query. Some measures, such as dynamic prevention by the integration of extra meta data, require less human interaction to enforce a limit on user input which also has the propensity to alter the semantics of the original code. However, in these techniques, there is a requirement for additional effort to separate the application code from the user input. Using an architecture based on proxy to dynamically recognise user's input from the application-generated query has a high detection rate as well as poses no requirement of access to the source code or database of the web application.

(c) **Man-in-the-Middle Attacks (MITM)** – MITM attacks is said to happen whenever an assaulter seek to intrude into a conversation that is ongoing with the goal of injecting false information that will help to gain access to sensitive shared information. Secure Socket Layer (SSL) offers security for applications based on the web. Secure Socket Layer employs TCP to build end-to-end services that are reliable and secured by the use of three fundamental protocols, namely, alert protocol, change cipher spec and handshake. It targets the provision of authentication, message integrity and confidentiality to the users, etc., by the use of digital signatures, certificates and cryptography. By a thorough configuration, SSL attacks, e.g. Wrapping attacks, MITM, XSS etc., may be eliminated.

(d) **Phishing Attacks** – Here, the attacker takes advantage of cloud service and compromises a web link and causes the redirection of the user to a false link, and by hacking, the user's account secures access to sensitive data. This attack can be eliminated by detecting spam pop-ups or emails, which may be carried out by the use of anti-spam tools.

(e) **Denial of Service (DOS) Attacks** – Here, the attacker seeks to deny authorised users of services by launching ICMP flooding, UDP flooding, SYN flooding attacks, etc. on the server. This attacker seeks to disable services or break the network provided by the server by continuously sending data packets across to the desired server without data packets, changing nodes or decrypting encrypted data. The network bandwidth is occupied by the data packets, which also consume the server's resources [34].

(f) **Distributed Denial of Service Attacks** – DDOS is a higher kind of denial of service attack as it relates to bombarding the target server with a vast bulk of packets from several networks which have been earlier manipulated so as to disable the target server's services while also creating more traffic than what is found in DOS in a manner that will render it difficult for the targeted server to handle the requests.

An intrusion detection system is a program adopted to gather network traffic and analyse them and then generate alarms or alerts regarding identified intrusions (any malicious activities or violations of security policies) for the system to carry out every necessary action. This intrusion prevention software has the capability to detect an intrusion detection system and can take possible steps towards preventing likely counterattacks.

(g) **Sniffer Attacks** – The malicious agent uses programs to launch these attacks to enable a host use an Ethernet network to capture flowing packets by the insertion of the network interface card (NIC) of the host into the malicious code. If the transferred information in these packets is not encrypted, these programs can easily compromise it. There are two techniques that are reportedly used to identify sniffer programs. First, address resolution protocol (ARP) identify sniffer attacks by relaying trap address resolution protocol packets which include false hardware addresses to a host that is suspicious. Then, on the ground of the response of address resolution protocol reply packets, decision is carried out whether or not a suspicious host is using a sniffer. Second, round-trip time (RTT) takes advantage of RTT evaluation of ICMP packet samples and thereafter makes a probabilistic decision by the use of a statistical model.

(h) **Google Hacking Attacks** – This is otherwise regarded as Google dorking and is a method of hacking which uses Google search engine to identify loopholes in the security configuration. Taking advantage of search queries, these hackers can pinpoint vulnerabilities in security and can then gather information about their desired targets [35].

(i) **Cookie Poisoning Attacks** – Here, the cookie's contents are manipulated in order to secure access to an unauthorised web page or application [1]. The cookie comprises sensitive credentials regarding the data of the user, and the moment the hacker secures access to these contents, he also invariably secures access to the content therein and may not carry out illicit activities.

(j) **Malware** – this being an acronym for malicious software, its threat and detection are the top two challenges in the CC environment [36]. The developers of malware are attempting/trying to interfere or impede with the integrity, confidentiality and/or the accessibility of data and the systems in which they are processed, transmitted and stored [37]. Also, statistics have shown, based on a study, that there was a notable increase in the total volume of malware between the period of 2011 and

2019 [37]. Types of malware which include worms, Trojans, spyware, rootkits, bots and backdoor are usually shared by hackers in a way that makes their variants have little differences [36]. According to Rains [37], one way to aid the detection of malware attacks is by being knowledgeable about it and use of anti-malware, realising that one malware can have multi-malware characteristics. Also, focusing on cybersecurity fundamentals will aid the mitigation of malware threats, and blocking access to Internet regions without legitimate business purposes can reduce exposure to malware.

(k) **CAPTCHA Breaking Attacks** – this is used to detect if the user is a human or a malicious program [1]. CAPTCHAs are, therefore, standard strategies for security which is employed to identify malicious software such as botnets, worms, Trojan, etc. The attacker may employ an audio system to break the CAPTCHAs and may use software for speech-to-text conversion to read the CAPTCHAs and can also break video-based and break image-based schemes. Letter overlap may be employed in order to prevent vertical segmentation attacks. Connected letters will render it very difficult for OCR to separate the words. Using various random alphabets and various fonts makes it difficult to break down CAPTCHAs. Moreover, it will be harder to break down CAPTCHA if the string length can be made long. The perturbative background can be made to have colours, lines, dots, rectangles, circles, etc., which will be quite hard to break down [38]. Some other attacks on cloud computing include attacks on domain name server (DNS), wrapping attacks, reused IP address attacks, zombie attacks and hypervisor attacks [1].

9.3 Malware Detection Methods in Cloud Computing Infrastructures

9.3.1 Overview of Malware in Cloud

The word "malware" is gotten from two words, "malicious and software". Therefore, malware is software which put harmful and malicious effects on the OS (operating system), software or other components. Malware is malicious software that is designed with the intention to cause harm or damage to its target system. These are creating huge challenges in today's technology world. There are various kinds of malware: Trojan horses, worms, backdoors, viruses, spyware, rootkits, ransomware and botnet, plus several other kinds of software with undesirable behaviour. Considering the exponential growth of the Internet, malware has also grown to be one of the primary cyber threats encountered today. Any program or product that executes malicious actions, including information theft, spying, etc., is malware. A highlight of the various families of these malware is given below:

Worm This is a program with features similar to the viruses but rather affects the network instead of the host machine. It is designed to infect another machine after reproducing itself [39]. They are spread across a computer network and depend on security failures for the penetration of their target machine. The majority of worms are designed to steal data, delete data and ultimately have them spread to other systems. Virus: The major characteristic of a virus is malware built by cybercriminals by infecting the target machine's file [39]. This sort of program self-replicates on the host machine and then connects to documents that eventually turn out to be their carriers. The design of a virus is such that it would spread from one machine to another.

Trojan horses This is a harmful code masquerade as legal software or useful code. Cybercriminals employ it to land access to the system of the user. It's a highly distinct malware type that appears useful at first glance, but has an embedded malicious code concealed and runs alongside when the program is run on the system. The provisions of social engineering are employed to deceive users and cause them to execute Trojans in their systems.

Rootkits These are a group of software tools built to provide an unsanctioned user with administrator privileges access to a system. Once the software has been installed, it may execute files or change system settings remotely [39]. They are advanced malware since they deal directly at the kernel level, a dangerous process that may result in the crash of the entire machine. They create direct harm to the victim's infrastructure. However, these programs cannot self-propagate but must be installed on a host system. Backdoors: These are the loopholes the cyber attacker uses the program to exploit, and these loopholes are created by the attacker to access to steal the victim's information.

Adware Unlike Trojan and many others, adware is not a direct harmful code but slows down the host machine's functioning by consistently displaying ads that land users on harmful pages or sources. Adware is targeted at displaying advertisements and redirecting search requests to other websites being advertised. Adware can explore functions like cookies to collect information about the user, e.g. the websites visited. Taking advantage of the information, customised ads are then displayed.

Ransomware This malicious code is relatively newer. It encrypts all the files or data of the user and then asks for ransom in order to decrypt the vital user's data. It also shows a warning alert requesting that Bitcoin be decrypted and the files or data be recovered.

Bots and botnets These are harmful codes designed to invade a computer and carry out instruction the moment it receives instruction from a remotely controlled server. Just like viruses and Trojans, bots can replicate itself. An array of bots described as botnets may be employed in launching DDoS (distributed denial of service) attacks to render the communication across a network temporarily inaccessible.

Key logger This malware tool takes a record of all the activities carried out on a monitoring tool similar to a machine. It regularly bypasses the permission of the user to execute. A key logger is predominantly utilised in obtaining confidential data, security phrases, passwords and usernames.

9.3.2 Overview of Anomaly Detection in Clouds

For several years, anomaly detection has remained an active area of research. Several techniques for various application domains and scenarios have been developed. The authors in [40] demonstrated through a survey across a number of disciplines the prediction, detection as well as estimated anomaly detection accuracy. Meanwhile, a thorough survey of the deployment of different schemes for anomaly detection in the area of IP backbone networks is found in the work of Marnerides in [41]. The production of the EbAT system a while ago gave room to the online analysis of a couple of metrics obtained from components at the system-level (such as memory utilisation, CPU utilisation on rack servers, operating system's read/write counts, etc.). The proposed system demonstrated potential in the aspect of monitoring scalability and detection accuracy; however, pragmatic cloud scenarios were not adequately emphasised in its evaluation [42]. The CP intrusion detection system proposed in [43] for detecting intrusions and attacks at different cloud layers was deficient for use in dynamic cloud environments because of the flexibility required. About a decade ago, Lee et al. [44] brought forward a multi-level approach, which offers rapid detection of anomalies found in the system logs of the operating system of each guest. One of the primary demerits of the approach is its lack of scalability since there is a requirement for higher resources under heavy system workload. Also, it is made to assort text-based log data, where the effects of the malware may not be manifested. In [45] the authors made a new prototype which supported an online spatiotemporal scheme of anomaly detection in a cloud setting. From there, the researchers had the ability to formulate and as well implement a wavelet based multi-scale system for anomaly detection. The system is anchored on measured cloud performance metrics such as memory or CPU utilisation, which is gathered by several components such as system, software and hardware within the institution-wide cloud setting under examination. The findings were promising as the proposed method attained a sensitivity of 93.3% in the detection of anomalous events, even as merely 6.1% of the entire events reported were false alarms. However, just before this time, the work of Pannu et al. in [46] brought in a framework of online adaptive anomaly detection that could identify failures by the analysis of runtime and execution metrics where the conventional two-class support vector machine (SVM) algorithm was used. In an actual practical employing above 362-node CC environment in a university setting, the findings revealed an efficient proposed system that had an overall sensitivity of 87% in anomaly detection. However, the primary challenge with this work was that the conceptualisation of the two-class algorithm of the support vector machine (SVM) suffered the problem of data imbalance [41], which impacted the training phase, and ultimately led to various mis-groupings of newly tested anomalies.

The work of Watson et al. in [42] on the online anomaly detection approach applicable at the cloud infrastructure's hypervisor level covered the area of early detection of an attack and confronted the algorithmic constraint usually acquired in a majority of the conventional two-class on n-class techniques based on machine learning (e.g. Bayesian classifiers, artificial neural networks, two-class SVMs) whenever they are applied to cloud environments. The work, which emulated "static" detection and even detection in VM "live" migration scenarios, was said to have stemmed from suggestions from cloud operators. By the exploration of features collected at the system and at a cloud node's network levels, it was demonstrated that the scheme could attain a high accuracy of detection of over 90% in the detection of various kinds of DoS attacks and malware. It was thereafter reported that the extracted features aided in detecting anomalies throughout the testing at a minimal time cost. Watson et al. in [42] recommended that this work can be furthered by expanding the feature reset to incorporate statistics obtained from the usage of vCPU and a more thorough investigation of process handles, which may help greatly in detecting very stealthy malware. However, this is expected to instigate a computational trade-off, considering that a more thorough introspection will demand more system resources.

9.3.3 Malware Detection in Cloud Infrastructures

In CC, there are three modules, and malware detection (and prevention) systems operate on these module's databases, virtual machines and networks for different attacks:

1. Malware Injection: This attack targets mainly servers and is designed such that whenever data is sent from the server, it contains embedded malicious code automatically, which will negatively impact all the server's clients.
2. MITM: These modes of attacks are utilised primarily for stealing users' data.
3. DDoS: These attacks are executed to disrupt the entire network by using useless traffic to jam the network traffic.

One of the most fundamental challenges with the development of secure cloud-oriented and resilient mechanisms is around the adequate detection and identification of malware. This is because, in the predominant of cases, malware is the initiation point for large-scale email spamming and phishing, DDoS attacks, primarily through botware deployment.

Malware detection and prevention systems (MDPS) help in the detection of malware by the use of signatures or several other heuristics techniques or other string-based or rule-based pattern machine. Antivirus are also typical examples of malware detectors. The majority of malware writers are accustomed to these methods used by antivirus, so they devise new means of evading these techniques by making modifications to the malware code, or they input junk data, which changes the file's hash and then renders it undetectable. These agents use password-protected approaches or encryption tools to escape detection. Conventional approaches for the detection of attacks on cloud infrastructures or the virtual machines they host are insufficient in addressing cloud-related issues, in spite of the great efforts put into previous studies as regard the behaviour of some kinds of malicious programs on the Internet [42].

Techniques for malware detection basically utilise two inputs for detection: (i) malware signature, rules or behaviour from the database and (ii) the target program to be evaluated for malicious intent. For higher security in the cloud, MDPS also employ real-time malware analysis. This real-time technique for prevention is very vital in dealing with the daily growing array of malware since it shields the user from unknown attacks and malware that may compromise the host machine and affect the user. The following is a detailed description of various approaches employed in malware detection in cloud infrastructures today.

9.3.3.1 General Malware Detection Approach

In the most basic malware detection, two methods are employed: shallow analysis and deep analysis. In the shallow method, the process parameters on the victim machine are checked. The check takes place prior to and following the malware execution. This is to diagnose the events the malware initiated and their effect on the machine. The following are the compromise to the machine instigated by malware: (i) update or change in key or windows registry entries; (ii) unexpected/unanticipated raise in the number of running processes on the system; (iii) file deletion, creation or modification. In the shallow analysis, these parameters mentioned above are explored to develop a profile of the machine. Different snapshots were done before the analysis and after the analysis, and both categories were compared to detect if there were unexpected changes that may be attributed to malware. The shallow analysis considers the following parameters: memory usage, CPU speed and usage, number of users, process state, and number of processes. However, in deep analysis, the file's hash is calculated, and the file is checked for malware patterns and strings, for accurate detection of malware.

9.3.3.2 Signature-Based Detection

This approach seeks to define a hash file set that can be employed in deciding that a particular pattern belongs to the malware. Because of this characteristic, the signature-based system may attain high detection speed since there's less size, accuracy as well as the minimal rate of false-positive results in the identification of intrusions in the system. The major challenge of this system is that if a minor modification is made to the file, it changes the entire signature, and the signature-based detection becomes inefficient. Thus, this technique may generate false-positive results or be unable to detect unknown attacks. Despite these drawbacks, signature-based detection is being employed due to the ease of updating and maintaining signatures. The signatures comprise various elements used in identifying the traffic. The parts of a signature include the header (e.g. ports, destination address, source address) and its options (e.g. metadata, payload), which are explored in determining if the network tallies with a known signature or not. Malware signature-based detection is adopted in cloud platforms for detecting samples that already exist in the database. The technique may be employed at a cloud gateway to detect external intrusions after a cloud firewall detects internal/external intrusion.

Signature Optimising Pattern Matching These methods depend on the signatures stored in the database. Here, a string matching algorithm is used for detecting intrusion. This is often employed in the analysis of DNA and different protein sequences. This approach is important for malware detection because of its provision of basic ground level for the detection of viruses. With the availability of a database of viruses, the method secures the system against almost all known types of malware. Whenever unknown malware samples are detected, there is a need for them to be added to the database. Therefore, when there's the detection of virus using the signature match, the signature of the virus is temporarily stored [32]. This is to ensure that other replicas have no need to match against the vast signatures in the actual database of signatures. This technique of pre-comparison will ensure that the signature matching times are drastically reduced.

9.3.3.3 Heuristic Detection

This method may be used in dealing with certain parts of the file where there is a maximum probability of finding the malware and thereafter calculating a hash of the specific byte of strings. Here, several signatures can be generated from the file. This is an advancement over the conventional signature-based approach since the whole file signature is estimated, which may not be effective when the signature changes when junk data is added or there's a modification to a little part of the code. But in this fuzzy logic, contrary to the calculation of the signature of the entire file, only a certain part of the signature is calculated, which will have a straightforward approach and will build more efficiency for the system. The method is also capable of detecting unknown samples of malicious code. It minimises the false-positive rates and increases the scanning speed. This method is also used for real-time detection of intrusion. The parameters to be compared are extracted from the network packet header. This method is employed in large-scale network attacks.

9.3.3.4 Automatic Signature Extraction

Traditional methods of signatures extraction depend mostly on the manual extraction by an expert of a sequence of bytes. The byte sequence is embedded in the unknown file's executable part, and there's a high degree of unlikeliness that these kinds of a string are seen in normal files. This process is carried out manually for the detection of a string of bytes, which is time-consuming and renders signature extraction quite tedious. Therefore, there's a need for a system that will be used to extract signatures from malware samples automatically.

One may use any byte sequence from a malware executable portion for the signature detection of that malware. Sometimes, the generated malware signatures can match some malicious contents and thus stimulate a high rate of false-positive results. Therefore, avoiding or reducing the rate of false-positive tests demands crucial consideration for the extraction of malware signatures. Another major factor to be considered is the time required to detect the malicious program in the network traffic. To reduce malware signatures' scanning time, there's a need to restrict the signature length. Also, the time to scan is reduced by the reduction of the signature of the malware and just those that may detect the majority of the samples of the malware. It is thus recognised that several malware samples comprise these popular executable parts. Since the concerned activity here is to reduce the total number of signatures and raise the efficiency of the signatures, there's a need for researchers to seek a minimum cover set of signatures that can effectively detect all the samples. This will bring about increased scanning speed, as these signatures have a high likeliness of detecting already existing malware variants. However, it is difficult to reduce the scanning time as well as the false-positive rate at the same time while still maintaining the program's efficiency and identifying the optimised set of malware signatures automatically [47].

9.3.3.5 *Anomaly-Based Intrusion Detection*

This malware detection method is as well referred to as behavioural-based detection. Here, the events are the major objects of monitoring at regular intervals, and the events are then covered for analysis. The analysis is carried out according to the behaviour change in the machine following the introduction of the malware. The events preceding and following the injection of malware are compared, and then the code may be declared malicious or not. Comparatively, unlike in the signature-based approach, this system can be used to detect unknown malware samples [48]. The key feature of this method is the efficient generation of rules that lowers the rate of false-positive results for both known and unknown attacks.

9.3.3.6 *Association Rule-Based IDS*

As malware and even virus samples are easily found on the Internet, malware attacks are quite popular now as one doesn't have to be an expert to launch a malware attack, as the code is readily available, and the attack may be launched after some modification. In order to detect such malware attacks, one may use the signature a priori algorithm, which sorts common subsets containing elements of original attacks of a particular attack set. This method employs signature-based algorithm for the generation of signatures for detection of misuse. Meanwhile, the primary challenge of this algorithm is the time taken in generating signatures. A scanning reduction algorithm was proposed for the reduction of the number of database scans for an efficient generation of signatures from formerly identified attacks. However, this raises the rate of false-positive results since there's a production of unwanted patterns. Cloud attacks can occur at different levels. There are four basic types of intrusion detection systems used in the cloud: network-based intrusion detection system (NIDS), host-based intrusion detection system (HIDS), distributed intrusion detection system (DIDS) and hypervisor-based intrusion detection system. Some studies have given various viewpoints to their address of cloud security (e.g. hypervisor, the network, operating system and guest VM) under different approaches that are derived from either statistical anomaly detection models or traditional rule-based intrusion detection systems (IDSs).

9.3.3.7 *Convolutional Neural Network (CNN)*

Convolutional neural network is a kind of deep learning that has been applied in the analysis and classification of images [49]. One comparative benefit of this method is that only little pre-processing is required, as opposed to similar algorithms for image classification, even as it works on unprocessed data. It serves as a feature extractor, a provision that is very convenient since most cases of feature selection requires the input of human experts. The test is carried out by executing different malware (especially rootkits and Trojans) on virtual machines. Abdelsalam introduced a detection method for virtual machines which engaged a 2D CNN model by taking advantage of the performance metrics. On the testing dataset, the findings demonstrated an appreciable accuracy of around 79%. The challenge of mislabelling was noted and improved on by the introduction of the 3D CNN model, which utilises samples over a specified time window. A 3rd dimension was added to the 2D input matrix. A large improvement was recorded, and an accuracy of around 90%, which is practically acceptable, for the 3d CNN 2 classifier was shown.

9.3.4 *Challenges of Malware Detection in CC Infrastructure*

Some of the challenges encountered in the area of malware detection in the cloud are presented; thus, in the area of data collection, only a few of the methods used give consideration to feature selection in the process of classification towards increasing result accuracy [50]. There is also the challenge of the continued surfacing of malware. In terms of analysis, many of the techniques, such as the anomaly-based approach, are restricted to a particular amount of malware. Also, many of the approaches are unable to unravel the topmost number of features required to train a classifier. In the area of response, there are a high number of false-negative and false-positive results. Also, there are scalability challenges in the handling of a massive number of malware samples. Also, there is the challenge of limited storage and computing resources. Also, the demand to gather new malware and continually make it benign is very tasking.

Some studies, such as Watson et al. in [42], criticised signature-based approaches such as intrusion detection systems (IDSs), on network packets, or the use of virtual machine introspection (VMI) for the detection of threats in a particular operating system of a virtual machine, saying that these purely signature-based approaches do not have the capability for a robust scheme that covers for future threats from novel strains of malware, as a result of its simplistic rule-based nature.

A number of techniques for anomaly detection are designed to reactively and proactively detect threats that are cloud-specific. Nevertheless, as a result of their complex statistical measures, they frequently demand prior knowledge and majorly lack scalability, therefore rendering them unfit for online detection in cloud infrastructures [45, 51].

9.4 Safeguarding Against Attacks in Cloud Infrastructure

This section addresses measures or tools that are either being implemented or have been proposed for implementation in safeguarding against attacks in cloud infrastructure. The primary challenge in cloud environments is the provision of security that touches the area of isolation and multi-tenancy, which guarantees the customers more confidence beyond just the "trust us" mantra of clouds. Nevertheless, a holistic approach is sacrosanct to achieve a comprehensive level of security. Securing the infrastructure at various levels, such as the application level, host level and network level, is important for the continuous running of the cloud. To protect the cloud against various malware and security threats, e.g. backdoor, bots and botnets, SQL injection, DoS and DDoS attacks, cross-site scripting (XSS) and forced and Google hacking, various cloud service providers may implement various security techniques. Different kinds of security breaches, as well as the implemented or proposed safeguards, are discussed in the following section.

9.4.1 Host Level Security

This describes how the server prevents threats or mitigates the impact of an invasive attack and what response is designed for emerging threats. In the evaluation of host security as well as risk management, attention is given to the delivery models for cloud service, e.g. IaaS, PaaS, PaaS and private, public and hybrid implementation models. However, the duties for host security in PaaS and SaaS services are committed to the provider of cloud infrastructure. The hosts that the cloud system supports are supported by the IaaS clients [52]. A vital technology-Web 2.0, which enables the utilisation of SaaS, takes away tasks such as installation and maintenance of software from users. With the increase in the use of Web 2.0, there's an urgent need for the environment now more than ever [53].

SQL Injection Attacks This involves the insertion of a malware code into a standard SQL code. By this, malicious persons secure unsanctioned access to the database and are then able to penetrate sensitive information [54]. Different techniques, such as preventing the use of SQL that is dynamically generated in the code or the use of filtering techniques, help in sanitising the input of the user, etc. and consequently help to mitigate SQL injection attacks. There has been a proposal for an architecture that is based on a proxy which dynamically detects and extracts the input of the users for SQL control sequences [52].

Cross-site scripting (XSS) attacks Since the introduction of Web 2.0., these attacks involving the injection of scripts into web contents (either by reflected XSS or stored XSS) have grown in popularity. We may classify a website as dynamic or static according to the kind of services provided. Dynamic websites, because of the multi-fold services they provide to users, are more vulnerable to attacks, e.g. XSS attacks, than static websites. Out of curiosity or even unknowingly, users click on pop-ups and links orchestrated by malicious programs, which allows the intruder to have control over the private information of the user, which may then be used to hack their accounts. Several techniques, such as active content filtering, technology for preventing content-based data leakage, for detecting web application vulnerability, have been developed to safeguard against XSS attacks. Also, an approach which is blueprint-based has been proposed that reduces the reliance on a web browser for the identification of untrusted content.

Man-in-the-middle attacks (MITM) In this kind of attack, a certain entity makes an attempt to create an intrusion into a conversation that is ongoing between a client and a sender in order to know of vital data being shared or to inject fake information. Different tools that implement hard encryption technologies, such as Ettercap, Cain, Dsniff, Airjack, Wsniff, etc., have been designed to safeguard against these attacks [52].

Also, test of data communication between license parties and proper SSL configuration has been found helpful in mitigating the risk of MITM attacks [55].

9.4.2 Network-Level Security

Several security threats are recorded in different types of networks, whether private or public, shared and non-shared, large or small area networks. However, different levels of security are faced by private and public clouds. The private cloud is said to be less vulnerable compared to the public cloud. This justifies the preference for private cloud by organisations for the migration of their sensitive data [52].

However, even in the public cloud, strong methods for network traffic encryption can be employed to safeguard against the breach of confidential data, including Transport Layer Security (TLS) and Secure Socket Layer (SSL). Also, there is a recommendation for the implementation of protection features like user authentication [56], data integrity, privacy protection, data security, virtual machines accessibility, web server security, recovery and compatibility. There's also a need for research efforts towards maintaining the consistent and smooth operations of the cloud [52]). The issues found in network-level security that demands preventive or management measures include sniffer attacks, DNS attacks, DDoS and DoS attacks, reused IP address, etc. [57].

Domain Name Server Attacks For easy remembrance, DNS helps to translate the domain name to an IP address. However, in some cases where after the user has called the server by name, the user is routed to a different malicious cloud different from the one they intended, where the use of IP addresses is therefore not feasible every time. Measures of reducing the impact of DNS threats include DNSSEC (Domain Name System Security Extensions). Nevertheless, there are reports of the inadequacy of these measures when a malicious connection is used to reroute the path between a receiver and a sender.

Sniffer Attacks These attacks are achieved using malicious software that is able to capture a network's packets, and without encryption of the data transferred across the packets, vital data can be traced, read or captured. A platform anchored on address resolution protocol (ARP) and round trip time (RTT) for malicious sniffing detection can be employed on the network for detecting running sniffing systems [52].

BGP Prefix Hijacking This kind of attack at the network level involves making a false announcement concerning the IP addresses that are linked to an autonomous system (AS). Therefore, malicious attackers are able to get into untraceable IP addresses. An AS can do an information broadcast of an IP within its regime across its entire neighbours [53].

9.4.3 Application Level Security

This security level describes the use of hardware and software resources to secure applications in a manner that malicious parties are unable to make illicit changes or exercise control over the applications. A lot of platforms are safeguarded at the network level, although certain application level issues can grant access to unsanctioned users [52]). In recent times, attacks are being camouflaged as trusted users, and the system addresses them as trusted users, which allows infiltration of victim platforms and stealthily corrupts the victim's entire data.

There is, therefore, urgency for the installation of a more sophisticated level of security checks to mitigate these risks. The conventional approaches for handling security challenges have involved the use of task-oriented ASIC devices that can address a certain task that offers higher security levels with high performance [58]. However, because of the adaptable and dynamic nature of application-level threats, these approaches have been found to act comparatively slow. The adaptability of open-ended systems, as well as the capabilities of a closed system, have been combined in developing security avenues that are anchored on Check Point Open Performance Architecture by the use of Quad-Core Intel Xeon Processors [53]. Moreso, in the virtual environment, institutions such as VMware use the technology of Intel Virtualization for better security bases and performance. The issues found in application level security which demands preventive or management measures include SQL injection attacks, cookie poisoning, CAPTCHA breaking, backdoor and debug options, DoS attacks, etc. [28].

Denial of Service (DoS) Attacks This kind of attack, besides creating congestion, raises the bandwidth being consumed and thereby renders particular parts inaccessible to the users of the cloud. The use of an intrusion detection system is a common technique for safeguarding against these attacks [57], and separate IDS is loaded in each cloud. Also, the defence federation has been employed for protecting the cloud against DoS attacks [59], although the extent of its use is not well documented. The various IDS systems function on the group of information exchange. A co-operative IDS also has the ability to alert the entire system on the occasion of an attack on a particular cloud.

DDoS Attacks Distributed denial of service functions like an advanced DoS attack in denying users of the availability of the server's services by bombarding the destination server with large packet numbers that overwhelm the ability of the server. However, unlike in DoS attacks, the attack in DDoS is distributed from different already compromised dynamic networks. The attackers then make certain information to be accessible at particular times. Therefore, the attacker has control over the type and amount of information available for consumption [58].

Cookie Poisoning This involves a change or modification of the cookie's content to gain unsanctioned access to a web page or application. The cloud infrastructure can be safeguarded against cookie poisoning by the implementation encryption scheme or by carrying out regular cleanup of cookie data.

CAPTCHA Breaking CAPTCHA itself is designed to prevent computers or bots from using the resources on the Internet, thereby preventing over-exploitation and spam of resources on the network by bots. CAPTCHA has suffered breaking from spammers who have developed strategies to defeat the test. A method of attack involving static OCR (optical character recognition) can be addressed by the use of CAPTCHA design principles of single-frame zero knowledge. Hardware and software tools can also be employed to protect applications against CAPTCHA breaking [52].

Rootkit in Hypervisor Concerns The overall concept of cloud computing is anchored primarily on the idea of virtualisation. In this virtual setting, a hypervisor allows a system to simultaneously run multiple OS and provide resources distinctly to each OS to avoid interference with one another. Various components of the hypervisor can be the subject of different kinds of attacks [53]. By majoring on an advanced comprehension of the behaviour of the different components of the hypervisor architecture, the use of intrusion prevention system and instruction detection system and firewall implementation, and inter-communication between the different components of the infrastructure, can be used to monitor the actions of the guest virtual machines [60].

Backdoor and Debug Options This attack is relatively passive, which affords the malicious party's penetration into compromised systems and, by the use of backdoor channels, gains access and control to the confidential data and resources of the victims and even turns it into a subject of DDoS attack. Advanced authentication and isolation methodologies between virtual machines can help to safeguard against these attacks.

9.5 Discussion and Analysis

This chapter aims to systematically analyse malware attacks and the techniques for safeguarding against malware challenges in cloud computing architecture. The research will also seek to make recommendations on the applicability of these techniques. Cloud computing is one of the decade's most trending discussions in information technology. It is described as a model for providing on-demand, convenient and ubiquitous network access to a shared pool of computing resources that can be configured (such as storage, networks, servers, services and applications) and may be provisioned rapidly and released with little interaction with the service provider or little management effort. The cloud itself is not an array of software, hardware or services but a vast shared resource that accommodates a large volume of users and offers dynamic access that is dependent on the demands. This in itself implies that cloud users have no need to possess the infrastructure enabling various computing services. But much more, the services become accessible to a computer from any location in the world.

This research found that some of the greatest selling points of cloud computing include its inherent features of multi-tenancy and, high scalability, enhanced flexibility, which transcends the provisions of previous computing methodologies. The provisions in cloud computing will enable the society to handle future challenges in quality assurance, information security and big data management. Furthermore, its possibilities of accessing new innovations like decentralised ledger technology, artificial intelligence, etc., as cloud computing services, are increasing. Despite all these benefits and potentials, the cloud is found to be highly vulnerable to a lot of security challenges, including malware. Thus, security is a major concern in cloud computing and a subject of discussion in several quarters. Therefore, research towards detecting this malware as well as safeguarding the cloud architecture against malware attacks is increasing.

As a fundamental objective of this research, security and malware threats in cloud computing architecture were analysed. The study identified that despite the huge benefits of cloud computing, it has met a growth of malicious activities, codes and programs targeted at the infrastructure. Reports that date back to over a decade ago have it that with the teeming adoption of cloud computing, one can expect to experience several security exploitations with cloud service providers as well as users, which will shift research focus to fixing these loopholes. In this study, some of the most commonly reported security threats identified in cloud computing (CC) infrastructure include data loss and breaches, malicious insiders, account or service hijacking, identity theft, phishing attacks, man-in-the-middle attacks, denial of service (DOS), distributed denial of service (DDOS) attacks, cookie poisoning attacks, wrapping attacks, etc. The study also recognised that as much as these all could employ different approaches and be launched against certain components of the infrastructure per time, several variants of malware are the reason for these attacks.

Therefore, there is a need for security at various levels towards the appropriate implementation of CC environment, e.g. Internet access security, server access security, program access security, data privacy security and database access security. Also, security needs to be guaranteed at the host, network and application layer for the smooth running of the cloud. Therefore, the issues around the security of cloud infrastructure are found in the description as well as the implementation of security implications that are provided by every relevant party in the process.

In this study, we found that with the exponential growth of the Internet, malware has also grown to be one of the primary cyber threats encountered today. The malware identified includes Trojan horses, worms, backdoors, viruses, spyware, rootkits, ransomware and botnet, plus several other kinds of software with undesirable behaviour. This study identified various malware detection techniques adopted in the cloud, e.g. malware detection and prevention systems (MDPS), antiviruses, use of virtual machine introspection (VMI), etc. The broad approaches used include automatic signature extraction, anomaly-based intrusion detection, association rule-based IDS and convolutional neural network (CNN). Nevertheless, these methods do not perform satisfactorily, perhaps because of the continued surfacing of malware. Also, in the aspect of data collection, very few measures employed consider feature selection in the classification that is aimed at enhancing the accuracy of the result. Also, strategies like the anomaly-based approach are limited to a certain number of malware. The systems encounter scalability issues in their attempts to handle big amount of malware samples. Also, because of the complex nature of anomaly detection, they frequently demand prior knowledge and majorly lack scalability, rendering them unfit for online detection in cloud infrastructures.

However, it was identified that in spite of all these measures, malware writers never cease to devise new means of evading these techniques by a change of file's hash, modifications to the malware code, etc. In spite of the great efforts put into previous studies as regards the behaviour of some kinds of malicious programs, conventional approaches for the detection of attacks on cloud infrastructures or the virtual machines they host are insufficient in addressing cloud-related issues. Even though new provisions have been laid down by the various malware detection and protection techniques, there are no methods with the ability to detect and safeguard against all sophisticated and new-generation malware. Organisations embracing cloud computing and enlarging their on-premise infrastructure must be abreast with the security burdens of cloud computing.

Also, beyond the detection of malware, to achieve a comprehensive level of security in the cloud, a holistic approach is sacrosanct. Various methods have been proposed for safeguarding the cloud against security attacks. However, this study found the safeguarding from the perspectives of application level, host level and network level to be more serviceable for the continuous running of the cloud. In the host level security, which involves paying attention to the delivery models for cloud service, various attacks such as SQL injection attacks, man-in-the-middle attacks (MITM) and cross-site scripting (XSS) attacks were identified. This study found such approaches as active content filtering, proper SSL configuration as well as data communication between license parties to be helpful in mitigating these attacks. At the network level, various kinds of threats identified include DNS attacks, sniffer attacks and BGP prefix hijacking. Measures such as Domain Name System Security Extensions, use of platform anchored on address resolution protocol (ARP) and round trip time (RTT) can be advantageous for malicious sniffing detection. Meanwhile, in application level security which involves the use of hardware and software for securing applications, the identified attacks include SQL injection attacks, cookie poisoning, CAPTCHA breaking, backdoor and debug options and DoS attacks. Several mitigation measures that can be employed in safeguarding against these attacks include the intrusion prevention system and intrusion detection system, regular cookie cleanup, firewall implementation and advanced authentication and isolation methodologies between virtual machines.

Although this study identified through the analytical study of several materials and previous works that there are several malware detection approaches as well as measures for safeguarding the cloud against various attacks, most of these measures are targeted at specific attacks or threats. This causes cloud service providers and several organisations and businesses with activities on the cloud to incur huge expenses in defending their systems against various forms of attacks they are vulnerable to. Also, these approaches will be incapable of handling multiple attacks simultaneously launched by attackers. But a generic/comprehensive framework is suggested to enhance the cost-performance ratio and offer multi-dimensional and multi-layer security. The primary criteria to be met in this comprehensive security framework are its interface with any cloud environment and its capacity to detect and address predefined and tailored security threats.

9.6 Conclusion and Recommendation

Considering the rate at which the cloud has taken over the information technology market, a more drastic shift to the cloud is anticipated in these coming years. When we consider the several benefits and potentials of cloud computing to organisations and individuals, it's succinct to say that cloud computing is a rapid revolutionary technology. Despite all the advantages, the cloud is found to be highly vulnerable to a lot of security challenges, including malware. Even the data in the cloud is vulnerable to various issues like integrity and confidentiality. To protect the cloud and utilise its full potential, there's a need to address these security challenges. This research is another development in the analysis of these security and malware attacks, with the potential to influence individual and corporate decisions in their adoption and maintenance of cloud computing infrastructure. The primary goal of this chapter is to systematically analyse malware attacks and the techniques for safeguarding against malware challenges in cloud computing architecture. Recommendations are also made concerning the applicability of these techniques. This research has analysed several security and malware issues found in cloud infrastructure, as well as the techniques for detection. The research also analysed measures for safeguarding cloud infrastructure against these attacks. The merits and demerits of some of these measures were also documented.

Even though new provisions have been laid down by the various malware detection and protection techniques, there are no methods with the ability to detect and safeguard against all sophisticated and new-generation malware. Organisations embracing cloud computing and enlarging their on-premise infrastructure must be abreast with the security burdens of cloud computing, as well as the available solutions for ensuring appreciable security in the cloud.

Cloud service providers and proactive organisations should invest in security and implement these measures in their infrastructure to ensure security and be able to explore cloud computing ahead of their enemies. Although the trends in malware generation and detection are dynamic and constantly changing, this research is a great guide for the activities of developers and computer scientists with security responsibilities in the cloud.

9.6.1 Recommendations

From our findings, various recommendations that may be considered for further studies or help to guide the decision and activities of stakeholders are presented below:

- There is a need to design a comprehensive framework for mitigating multiple malware and other security attacks in the cloud. This framework will be able to interface with any kind of cloud environment and have the capacity to detect and address predefined and tailored security threats. This will be a cost-effective approach and will help to secure the system against attackers who launch multiple attacks simultaneously, which would otherwise have overwhelmed the cloud and hurt the services provided.
- Since there is yet no omnibus approach to all security challenges, this study recommends that approaches that implement the combination of multiple detectors and/or mitigation approaches can be considered.

- In order to protect the cloud against external threats, there is a need for regular auditing of the cloud.
- Also, cloud service providers need to ascertain that all the service-level agreements are met, while human errors are appreciably minimised.
- Since several of the existing detection and prevention approaches require prior knowledge, there is a need for stakeholders, engineers, etc., to build sophisticated skills and competencies that put them ahead of the attackers.

References

1. Amara N, Zhiqui H, Ali A (2017) Cloud computing security threats and attacks with their mitigation techniques. In: 2017 International Conference on Cyber-Enabled Distributed Computing and Knowledge Discovery (CyberC). IEEE, pp 244–251
2. Aslan V, Ozkan-Okay M, Gupta D (2021) Intelligent behavior-based malware detection system on cloud computing environment. IEEE Access 9:83,252–83,271
3. Aslan Ö, Samet R (2017) Investigation of possibilities to detect malware using existing tools. In: 2017 IEEE/ACS 14th International Conference on Computer Systems and Applications (AICCSA). IEEE, pp 1277–1284
4. Aslan ÖA, Samet R (2020) A comprehensive review on malware detection approaches. IEEE Access 8:6249–6271
5. Ferrag MA, Friha O, Maglaras L, Janicke H, Shu L (2021) Federated deep learning for cyber security in the internet of things: concepts, applications, and experimental analysis. IEEE Access 9:138,509–138,542
6. Ferrag MA, Maglaras L, Moschoyiannis S, Janicke H (2020) Deep learning for cyber security intrusion detection: approaches, datasets, and comparative study. J Inf Secur Appl 50:102419
7. Mazumdar A, Alharahsheh H (2019) Insights of trends and developments in cloud computing. South Asian Res J Eng Tech 1(3):98–107
8. NIST Cloud Computing Security Working Group et al (2013) NIST cloud computing security reference architecture, National Institute of Standards and Technology, Technical Report

9. Golightly L, Chang V, Xu QA, Gao X, Liu BS (2022) Adoption of cloud computing as innovation in the organization. Int J Eng Bus Manag 14:18479790221093992
10. Zhao L, Zhou Y, Lu H, Fujita H (2019) Parallel computing method of deep belief networks and its application to traffic flow prediction. Knowl-Based Syst 163:972–987
11. Bullynck M (2018) What is an operating system? a historical investigation (1954–1964). In: Reflections on programming systems. Springer, pp 49–79
12. Qarkaxhija J (2020) Using cloud computing as an infrastructure case study-microsoft azure
13. Balatinac I, Radosevic I (2014) Architecting for the cloud
14. Cook A, Smith RG, Maglaras L, Janicke H (2017) SCIPS: using experiential learning to raise cyber situational awareness in industrial control system. Int J Cyber Warfare Terrorism (IJCWT) 7(2):1–15
15. Pathak N, Anwar S, Singhal V, Sharma N, Shukla AK (2019) Trends and augmentations of cloud computing. In: 2019 6th International Conference on Computing for Sustainable Global Development (INDIACom). IEEE, pp 373–377
16. Tang JG (2014) The research on cloud computing security model and countermeasures. In: Applied mechanics and materials, vol 511. Trans Tech Publications, pp 1196–1200
17. Zhang N (2021) An overview of advantages and security challenges of cloud computing. Int J Comput Sci Mob Comput 10(1):76–85
18. Sen S, Chaki R (2011) Handling write lock assignment in cloud computing environment. In: Computer information systems–analysis and technologies. Springer, Berlin, pp 221–230
19. Dasari Y, Dondeti V, Kalluri HK (2022) An effective framework for ensuring data privacy in private cloud. In: Smart data intelligence. Springer, Singapore, pp 535–547
20. Compastié M, Badonnel R, Festor O, He R (2020) From virtualization security issues to cloud protection opportunities: an in-depth analysis of system virtualization models. Comput Secur 97:101905
21. Deb M, Choudhury A (2021) Hybrid cloud: a new paradigm in cloud computing. In: Machine learning techniques and analytics for cloud security, pp 1–23

22. Maglaras L, Shu L, Maglaras A, Jiang J, Janicke H, Katsaros D, Cruz TJ (2018) Industrial internet of things (IIot). Mob Netw Appl 23(4):806–808
23. Bennasar H, Essaaidi M, Bendahmane A, Ben-Othman J (2021) A systematic literature review of cloud computing cybersecurity. In: Adv Dyn Syst Appl 16(2):1883–1919
24. Wei L, Zhu H, Cao Z, Dong X, Jia W, Chen Y, Vasilakos AV (2014) Security and privacy for storage and computation in cloud computing. Inf Sci 258:371–386
25. Shaar F, Ahmet E (2018) DDoS attacks and impacts on various cloud computing components. Int J Inf Secur Sci 7(1):26–48
26. Mladenov V, Mainka C, Schwenk J (2015) On the security of modern single sign-on protocols: second-order vulnerabilities in openid connect. arXiv preprint arXiv:1508.04324
27. Ramgovind S, Eloff MM, Smith E (2010) The management of security in cloud computing. In: 2010 Information Security for South Africa. IEEE, pp 1–7
28. Cook A, Robinson M, Ferrag MA, Maglaras LA, He Y, Jones K, Janicke H (2018) Internet of cloud: security and privacy issues. In: Cloud computing for optimization: foundations, applications, and challenges. Springer, pp 271–301
29. Hussin H, Salleh NA, Suhaimi MA, Ali AM (2019) Cloud computing practices and perceived benefits by SMEs in Malaysia: some empirical evidence. J Inf Syst Digit Tech 1(2):1–15
30. Suryateja P (2018) Threats and vulnerabilities of cloud computing: a review. Int J Comput Sci Eng 6(3):297–302
31. Vishnoi A, Mishra P, Negi C, Peddoju SK (2021) Android malware detection techniques in traditional and cloud computing platforms: a state-of-the-art survey. Int J Cloud Appl Comput (IJCAC) 11(4):113–135
32. Abdelsalam M, Krishnan R, Sandhu R (2019) Online malware detection in cloud auto-scaling systems using shallow convolutional neural networks. In: IFIP Annual Conference on Data and Applications Security and Privacy. Springer, pp 381–397
33. Papaspirou V, Maglaras L, Ferrag MA (2021) A tutorial on cross-site scripting attack: defense against online social networks. In:

Securing social networks in cyberspace, CRC Press, Boca Raton, FL, pp 277–296
34. Jensen M, Schwenk J, Gruschka N, Iacono LL (2009) On technical security issues in cloud computing. In: 2009 IEEE international conference on cloud computing. IEEE, pp 109–116
35. Jangjou M, Sohrabi MK (2022) A comprehensive survey on security challenges in different network layers in cloud computing. Arch Comput Methods Eng 29(1):1–22
36. Gao X, Hu C, Shan C, Liu B, Niu Z, Xie H (2020) Malware classification for the cloud via semi-supervised transfer learning. J Inf Secur Appl 55:102661
37. Rains T (2020) Cybersecurity threats, malware trends, and strategies: learn to mitigate exploits, malware, phishing, and other social engineering attacks. Packt Publishing Ltd, Birmingham, UK
38. Jeng AB, Tseng C-C, Tseng D-F, Wang J-C (2010) A study of captcha and its application to user authentication. In: International Conference on Computational Collective Intelligence. Springer, pp 433–440
39. Fui NLY, Asmawi A, Hussin M (2020) A dynamic malware detection in cloud platform. Int J Diff Equ (IJDE) 15(2):243–258
40. Dewa Z, Maglaras LA (2016) Data mining and intrusion detection systems. Int J Adv Comput Sci Appl 7(1):62–71
41. Marnerides AK, Malinowski S, Morla R, Kim HS (2015) Fault diagnosis in dsl networks using support vector machines. Comput Commun 62:72–84
42. Watson MR, Marnerides AK, Mauthe A, Hutchison D et al (2015) Malware detection in cloud computing infrastructures. IEEE Trans Dependable Secure Comput 13(2):192–205
43. Guan Y, Bao J (2009) A CP intrusion detection strategy on cloud computing. In: Proceedings. The 2009 International Symposium on Web Information Systems and Applications (WISA 2009). Citeseer, p 84
44. Lee J-H, Park M-W, Eom J-H, Chung T-M (2011) Multi-level intrusion detection system and log management in cloud computing. In: 13th International Conference on Advanced Communication Technology (ICACT2011). IEEE, pp 552–555

45. Guan Q, Fu S (2013) Adaptive anomaly identification by exploring metric subspace in cloud computing infrastructures. In: 2013 IEEE 32nd International Symposium on Reliable Distributed Systems. IEEE, pp 205–214
46. Pannu HS, Liu J, Fu S (2012) AAD: adaptive anomaly detection system for cloud computing infrastructures. In: 2012 IEEE 31st Symposium on Reliable Distributed Systems. IEEE, pp 396–397
47. Choo K-KR, Rana OF, Rajarajan M (2017) Cloud security engineering: theory, practice and future research. IEEE Trans Cloud Comput 5(3):372–374
48. Ahmim A, Ferrag MA, Maglaras L, Derdour M, Janicke H (2020) A detailed analysis of using supervised machine learning for intrusion detection. In: Strategic innovative marketing and tourism. Springer, pp 629–639
49. Abdelsalam M, Krishnan R, Huang Y, Sandhu R (2018) Malware detection in cloud infrastructures using convolutional neural networks. In: 2018 IEEE 11th International conference on cloud computing (CLOUD). IEEE, pp 162–169
50. Naseer M, Rusdi JF, Shanono NM, Salam S, Muslim ZB, Abu NA, Abadi I (2021) Malware detection: issues and challenges. J Phys: Conf Ser 1807(1):012011. IOP Publishing
51. Guan Q, Fu S, DeBardeleben N, Blanchard S (2013) Exploring time and frequency domains for accurate and automated anomaly detection in cloud computing systems. In: 2013 IEEE 19th Pacific Rim International Symposium on Dependable Computing. IEEE, pp 196–205
52. Alenezi M (2021) Safeguarding cloud computing infrastructure: a security analysis. Comput Syst Sci Eng 37(2):159–167
53. Bhadauria R, Sanyal S (2012) Survey on security issues in cloud computing and associated mitigation techniques. arXiv preprint arXiv:1204.0764
54. Nasereddin M, ALKhamaiseh A, Qasaimeh M, Al-Qassas R (2021) A systematic review of detection and prevention techniques of sql injection attacks. Inf Secur J: Glob Perspect 32(4):1–14
55. Munir K, Palaniappan S (2013) Secure cloud architecture. Adv Comput 4(1):9

56. Papaspirou V, Maglaras L, Ferrag MA, Kantzavelou I, Janicke H, Douligeris C (2021) A novel two-factor honeytoken authentication mechanism. In: 2021 International Conference on Computer Communications and Networks (ICCCN). IEEE, pp 1–7
57. Singh G, Sharma A, Lehal MS (2011) Security apprehensions in different regions of cloud captious grounds. Int J Netw Secur Appl 3(4):48–57
58. Lua R, Yow KC (2011) Mitigating ddos attacks with transparent and intelligent fast-flux swarm network. IEEE Netw 25(4):28–33
59. Lo C-C, Huang C-C, Ku J (2010) A cooperative intrusion detection system framework for cloud computing networks. In: 2010 39th International Conference on Parallel Processing Workshops. IEEE, pp 280–284
60. Lombardi F, Di Pietro R (2011) Secure virtualization for cloud computing. J Netw Comput Appl 34(4):1113–1122

Index

I. Almomani et al. (eds.), *Cyber Malware*, Security Informatics and Law Enforcement, https://doi.org/10.1007/978-3-031-34969-0

9783031349683